New trade theory, new growth theory and the developing literature on national systems of innovation have combined to produce an explosion of interest over the past two decades around issues of trade, growth and technical change. A similar focus has dominated public policy debates. This book represents a major contribution to such debates, focussing as it does on the interconnections between technology, on the one hand, and economic growth and international trade, on the other. It identifies and explains the links between these various processes and explores the effects of technical change on both growth and trade.

This book has clear policy relevance: market forces alone are seen to be far from adequate for the efficient generation, transfer and diffusion of innovation, while public policy is identified as playing a key role for a truly innovative economy.

Trade, growth and technical change

Trade, growth and technical change

Edited by
DANIELE ARCHIBUGI
and
JONATHAN MICHIE

CAMBRIDGE
UNIVERSITY PRESS

University Printing House, Cambridge CB2 8BS, United Kingdom

Cambridge University Press is part of the University of Cambridge.

It furthers the University's mission by disseminating knowledge in the pursuit of education, learning and research at the highest international levels of excellence.

www.cambridge.org
Information on this title: www.cambridge.org/9780521556415

© Cambridge University Press 1998

First published 1998

A catalogue record for this publication is available from the British Library

ISBN 978-0-521-55393-3 Hardback
ISBN 978-0-521-55641-5 Paperback

Cambridge University Press has no responsibility for the persistence or accuracy of URLs for external or third-party internet websites referred to in this publication, and does not guarantee that any content on such websites is, or will remain, accurate or appropriate.

Contents

List of contributors   ix
Foreword, by NATHAN ROSENBERG   xi
Acknowledgements   xiii
List of abbreviations   xv

1   Trade, growth, and technical change: what are the issues?   1
   DANIELE ARCHIBUGI AND JONATHAN MICHIE

2   The economics of technical change   16
   CHRISTOPHER FREEMAN

3   Uneven (and divergent) technological accumulation
   among advanced countries: evidence and a framework
   of explanation   55
   PARIMAL PATEL AND KEITH PAVITT

4   Technology and growth in OECD countries, 1970–1990   83
   MARIO PIANTA

5   R&D activity and cross-country growth comparisons   98
   MAURY GITTLEMAN AND EDWARD N. WOLFF

6   Aggregate convergence and sectoral specialisation in
   innovation: evidence for industrial countries   122
   DANIELE ARCHIBUGI AND MARIO PIANTA

7   International patterns of technological accumulation and
   trade   141
   GIOVANNI AMENDOLA, PAOLO GUERRIERI
   AND PIER CARLO PADOAN

8   Trade on high-technology markets and patent statistics –
     leading-edge versus high-level technology   168
     HARIOLF GRUPP AND GUNNAR MÜNT

9   High-technology industries and international competition   188
     PAOLO GUERRIERI AND CARLO MILANA

10   User–producer interaction, learning and comparative
     advantage   208
     JAN FAGERBERG

     References   226
     Index   268

Contributors

GIOVANNI AMENDOLA	Stet, Rome
DANIELE ARCHIBUGI	Institute for Studies on Scientific Research, Italian National Research Council
JAN FAGERBERG	ESST, University of Oslo
CHRISTOPHER FREEMAN	Science Policy Research Unit and MERIT
MAURY GITTLEMAN	US Bureau of Labour Statistics
HARIOLF GRUPP	Fraunhofer Institute for Systems and Innovation Research, Karlsruhe
PAOLO GUERRIERI	University of Rome 'La Sapienza'
JONATHAN MICHIE	Birkbeck College, University of London
CARLO MILANA	ISPE, Rome
GUNNAR MÜNT	Fraunhofer Institute for Systems and Innovation Research, BASF, Lundwigshafen
PIER CARLO PADOAN	University of Rome 'La Sapienza' and College of Europe
PARIMAL PATEL	Science Policy Research Unit, University of Sussex
KEITH PAVITT	Science Policy Research Unit, University of Sussex
MARIO PIANTA	Institute for Studies on Scientific Research, Italian National Research Council
NATHAN ROSENBERG	Department of Economics, Stanford University
EDWARD N. WOLFF	Department of Economics, New York University

Foreword

The contents of this book are especially timely and welcome. The last fifteen years or so have witnessed a drastic rethinking of the role played by technical change in the other two subjects that make up the book's title: trade and growth. It had long been apparent that technical change played a more powerful role in shaping both international trade and economic growth than was recognised in the formal literature of economics. However, theoretical advances in the 1980s and 1990s have had, as central and welcome features, an explicit recognition that technical change was a key to a deeper understanding of what goes on in each of these two critical areas. Formal model builders have been hard at work in these two realms that are now usually referred to as the new trade theory and the new growth theory.

New trade theorists have been spelling out the implications of the view that patterns of trade are strongly driven by (persistent) international differences in production functions; patterns of world trade are likely to be very different when countries do not have access to the same technologies. Indeed, it is now acknowledged (e.g., in the writings of Paul Krugman) that technical change in a world of uneven technical capabilities may turn out to be less benign than the conventional wisdom had once supposed.

A considerable accomplishment of the new growth theory has been to demonstrate that one can develop growth models in which technical change can be treated as an endogenous variable. Although the endogeneity of technical change has been recognised at least as far back as the writings of Marx, the ability to express this relationship in a formal model has been a powerful force in stimulating professional awareness and concern with this and related issues.

These theoretical achievements should be cause for considerable rejoicing. For one thing, no longer are economists who remain concerned with issues involving technical change regarded as indulging some merely arcane interest. In Chris Freeman's memorable expression, they are not 'rogue elephants' of the economic profession, 'whose work, although certainly of interest, should not be taken too seriously'. Rather, there is a serious sense in which it may be said that we are all Schumpeterians, or at least neo-Schumpeterians, now.

Thus, the intellectual case for 'getting down into the trenches' to do the hard empirical work, of the kind contained in this volume, while always strong, is now even stronger. Recent theoretical progress has increased the demand for the highly complementary inputs of empirical and historical research. It is in this context that the chapters in this volume have a compelling claim to the reader's attention. Each of them delivers empirical evidence of how technical change has been shaping the patterns of trade and growth among nations. As these chapters abundantly demonstrate, technical change covers a highly diversified range of activities when one looks across industry boundaries, and technical change as a learning experience is subject to a wide range of institutional and policy influences when one looks across national boundaries.

In making these observations I do not mean to suggest that good empirical work is to be valued solely for its role in providing inputs to the high theorist, although that is indeed a most valuable contribution. Beyond that contribution, such research is invaluable in its own right because there are, necessarily, sharply defined limits to the real understanding of technical change as an economic phenomenon so long as one remains at high levels of abstraction or aggregation. Good theory, if it is really to be good theory, has to be parsimonious. But good empirical research ultimately determines what ought to be *regarded* as good theory, or bad theory, and in the latter case it should also point the way to better theory.

A distinctive feature of the world economy, as these studies indicate, has been the increasing ease and speed with which it is now possible to move new technologies across national boundaries – an increasing speed towards which technology itself has played a major role, in the form of such innovations as jet aircraft, internet and fax machines. In this changing international context, issues such as technological convergence, new patterns of international specialisation, the altered nature of technological learning and the factors influencing the diffusion of that learning, all assume an increasing and more urgent prominence. It is a great virtue of this book that it not only identifies the forces at work; it also offers the reader a valuable set of new perspectives on their economic impact.

Nathan Rosenberg

Acknowledgements

This book developed from a special issue of the *Cambridge Journal of Economics* which we edited on 'Technology and innovation', published in February 1995. We selected the papers which focussed particularly on issues of trade, growth and technical change, and added additional pieces which we considered key to the development of research in this area.

We are therefore grateful to the Editors of the *Cambridge Journal of Economics* for having supported this project, and in particular to the Managing Editor Ann Newton for the additional burden of work it put on her. Geoff Harcourt, Alan Hughes and Jochen Runde provided helpful comments on the draft of our Introduction to that special issue for which we are grateful and on which we have drawn in our introductory chapter to this book. For comments on the draft version of that chapter we are grateful to Rinaldo Evangelista, Mario Pianta and Giorgio Sirilli.

Daniele Archibugi is grateful to Cambridge University's Department of Applied Economics, the Judge Institute of Management Studies and Robinson College, all of which provided hospitality during his two year stay while working on this project. We are also grateful to the European Commission for a 'Euroconference' grant awarded to us along with Jeremy Howells for 1995–7 which allowed us to draw on a range of expertise during three conferences at which the issues covered in this book were discussed in detail.

We are grateful to the authors for having delivered their chapters not only promptly but also on compatible and virus-free disks. And we are grateful to Nathan Rosenberg for having contributed the Foreword.

Patrick McCartan and Sophie Noël of Cambridge University Press did an excellent job throughout the whole process, from the production of the original proposals for this book, along with the companion volume on *Technology, Globalisation and Economic Performance* which we also edited and which was published by Cambridge University Press earlier this year, right through to publication. We are also grateful to Helen Dagut, Carol Jones, Robin May, Cinzia Spaziani and Patrizia Principessa for their help in the preparation of the manuscript.

Finally we are grateful, for having put up with weekend editing, respectively to Paola, seven-year old Clara and two-year old Orlando; and to Carolyn, seven-year old Alex and one-year old Duncan.

Daniele Archibugi
Jonathan Michie

Abbreviations

AG-TRAP	Arbeitsgruppe Transformationsprozesse der Max-Planck-Gesellschaft
CEPR	Centre for Economic Policy Research
CERGE-EI	Center for Economic Research and Graduate Education, Charles University, Economic Institute of the Academy of Science of the Czech Republik
CEU	Commission for the European Union
EBRD	European Bank for Reconstruction and Development
ECO	Economic Co-operation Organization
ECPR	European Consortium for Political Research
EECR	East European Constitutional Review
EUI	European University Institute, Florence
ILO-CEET	Internation Labour Office/Central East European Team
IMF	International Monetary Fund
IPSA	International Political Science Association
iw-trends	Institut der deutschen Wirtschaft – Trends
NBER	National Bureau of Economic Research
OECD	Organisation for Economic Co-operation and Development
RFE/RL	Radio Free Europe/Radio Liberty
SSRC/ACLS	Social Science Research Council/American Council of Learned Society
UCEMET	University Council for Economic and Management Education Transfer
UN	United Nations
UNDP	United Nations/Development Programme
UNICEF	United Nations International Children's Emergency Fund
UNICEF-ICDC	UNICEF – International Child Development Centre

WIIW   Wiener Institut für Internationale Wirtschaftsvergleiche
WZB   Wissenschaftszentrum Berlin
ZERB   Zentrum für Europäische Rechtspolitik an der
         Universität Bremen

1 Trade, growth and technical change: what are the issues?

DANIELE ARCHIBUGI
AND JONATHAN MICHIE

1 The revival of the economics of technical change

A survey of the main economic journals from the early 1950s to the late 1970s would show few articles on the theme of technical change. Moreover, the studies that were published were mainly concerned with the *impact* of technical change on variables such as growth, productivity, employment and competitiveness; much less attention was devoted to understanding the *sources* and *determinants* of the generation of innovations. Over the same period the world economy experienced its Golden Age, involving the massive and systematic exploitation of scientific discoveries and technological innovation. But, despite a few noteworthy exceptions, economists were unable to understand – or possibly just uninterested in – its sources. Hence Joan Robinson's remark that economists still treated technology as if it was provided by God and the engineers.

Over the last two decades, economists have to some extent addressed this gap. Articles on technical change are published frequently in mainstream journals and new specialised ones have been established. New data sources have been created and conferences on this issue are held with increasing frequency. Even politicians, who have for long preferred to rely on the advice of natural scientists and engineers, have allowed the advice of economists to inform their science and technology policies. While it was largely heterodox economists – Marxist, Schumpeterian, evolutionary or institutional – who first analysed technical change, orthodox economists have also increasingly turned to the study of the determinants of innovation.

Chris Freeman (this volume) provides a state of the art overview of the economics of technical change focussing especially on the latest developments. Of course, no single survey could cover the whole range of research carried out in this field nor do justice to all contributors and insights; indeed, this fact itself indicates the geometric rate of expansion in this field over the last years. A comprehensive reading list would need to combine the

1

huge number of references cited in each of the recent handbooks by Dogdson and Rothwell, eds. (1994), Stoneman, ed. (1995) and Freeman and Soete (1997), plus a variety of other surveys. In spite of this renewed interest, our knowledge of the determinants of innovation and its impact on the economy is still limited. This is due not to a lack of interest, but rather to the multifarious nature of technology. Technological change is related to a wide range of other economic and social phenomena, with inventive and innovative activities involving a variety of phenomena which are difficult to conceptualise and measure in simple models.

The complexity of the phenomenon of technical change is well illustrated by the following chapters of this book. The authors generally draw upon the eclectic approach which has its milestones in the writings of Adam Smith, Friedrich List, Karl Marx and Joseph Schumpeter and which has been reinvigorated in recent times by the writings of Nicholas Kaldor, Arthur Lewis and Albert Hirschmann. Chris Freeman, Richard Nelson, Nathan Rosenberg and their followers have explored in depth the implications of this approach for understanding technical change. Authors like Zvi Griliches, Edwin Mansfield and Mike Scherer in the United States and Keith Pavitt, Stan Metcalfe, Bengt-Åke Lundvall, Luc Soete and Giovanni Dosi in Europe have addressed new questions drawing on – and themselves generating – a wealth of new empirical research. This broad literature has been characterised by an openness to inputs from related disciplines. It has drawn from historians such as Moses Abramovitz, Alexander Gerschenkron and David Landes; from students of the firm such as Alfred Chandler, John Dunning and Edith Penrose; and has often been inspired by social theorists like Derek de Solla Price, Michael Polanyi and Thomas Kuhn. Although this literature has often been labelled 'evolutionary', the term 'institutional' may be more appropriate given the nature and variety of the contributions which this body of work now incorporates.

2 The links between technical change, growth and trade

The chapters in this book focus on the interconnections between technology, on the one hand, and the two major phenomena of economic growth and international trade, on the other. Historically, the dramatic rate of scientific and technical progress has taken place alongside two other epoch-making phenomena: economic growth and social and economic globalisation. Indeed, technological progress, growth, and globalisation describe the three most significant aspects of the long-term evolution of the world capitalist economy. These complex phenomena are obviously interrelated, although this does not mean that the linkages can be easily specified. On the contrary, the complex relationships between technical change, growth and trade are still subject to debate and controversy despite the large body of theoretical and empirical research which does exist. The chapters of this book are devoted to identifying and explaining some of these links. In particular, they are devoted to exploring the effects of technical change on both growth and trade.[1]

With the exception of Chris Freeman's chapter which critically surveys the field, the contributors to this book combine theoretical discussion with empirical research. This is no accident: the authors clearly take the view that the processes of trade, growth and technical change need to be documented using quantitative evidence. Theoretical models alone are insufficient; they need to be confronted with just such historical and empirical documentation. To a large extent, then, the contributors to this volume take the view that the deductive method in economics has major limitations when not nurtured by observation of economic and social life. Some of the best economic theory in this field has indeed emerged from the exploration of the industrial world.

There is another, although less explicit, element which unifies the studies here collected: all of them have clear policy relevance. The chapters have an implicit and sometimes explicit interest in understanding the devices which can be used to foster technical change. Market forces alone are seen to be far from adequate for the efficient generation, transfer and diffusion of innovation. Governments have various fundamental and non-replaceable roles in the process of promoting technical change which can take various forms: (a) the direct pursuit of scientific and technological activities, as in the case of universities and other publicly funded research institutions; (b) the financial support of innovation carried out in the business sector; (c) the supply of the necessary productive infrastructures, including education and training, standards and norms, and a legal system of intellectual property rights, to allow individuals and firms to innovate.[2]

The purpose of this introductory chapter is to summarise the main issues analysed in the subsequent chapters, relating them to the existing debates in the academic literature.[3] The next section highlights the main features of the key concept of this volume: technology. Sections 4, 5 and 6 discuss the relationship between technology and growth, while sections 7, 8 and 9 are devoted to the relationship between technology and trade.

3 The nature of technology

The contributors to this volume share a common view about the nature of technology. In this view, technology is seen as a multifarious human activity. Its variety makes it difficult to capture in aggregate concepts and measures such as those used in macroeconomic modelling. In the following chapters it is stressed that to understand changes in technology it is necessary to pay attention precisely to this diversity. What might be described broadly as institutional theory has made several attempts to identify the key aspects underlying the very complex process of technological innovation. The central conclusion might be summarised within the following four points.

The first is that technology is often proprietary in nature (Nelson, 1992a). Mainstream economics has tended to view technology as a public good, freely available to all economic agents, costly to generate but able to be assimilated with no or negligible costs.[4] The institutional approach rejects this assumption and argues that the producers of new knowledge have a variety of legal and economic methods to secure returns from their innovations. Certainly, would-be imitators can acquire technological competence but this is a costly, time-consuming process so that there will inevitably be some uncertainty about whether the economic returns obtained will repay the costs of imitating the innovation (see Mansfield, 1985).

Second, institutional theory points out that only a part of knowledge is codifiable in handbooks, blueprints, patents, scientific articles, etc. There is an equally important part of knowledge which is tacit and which can only be acquired by long processes of learning (Lundvall, 1996). In this framework, knowledge is specific to economic agents such as individuals, firms, industries and nations. Although some parts of know-how can be easily transferred from the producers to the users, this cannot be assumed to be a generalisable assumption.

Third, there are fundamental variations across different technological fields. While technological innovation in some fields may be relatively easily accessible to many agents, the expertise needed to have access to technological innovation in other fields will be confined to rather restricted groups of experts. Each technology system (semi-conductors, antibiotics and so on), industry and country has a specific regime of technological appropriation which makes the innovations either more freely available or else more proprietary in nature (Levin et al., 1987).

Fourth, the evolution of knowledge is highly path-dependent, that is, it is influenced by the knowledge already accumulated by economic agents in the past (Pavitt, 1988a). Although some agents might jump from one mix of competence to another, in the majority of cases, changes are rooted in the competences already acquired in the past (Nelson and Winter, 1977; David, 1985).

Such a conception of technology has important implications for understanding the links between technical change, growth and trade. The following sections are devoted to highlighting some of these implications.

4 The engine of economic growth

Technical change is one of the most important sources of long-term economic growth. New processes allow an increase in output per unit of input while new products create new markets and provide scope for output growth. Classical, neo-classical, Keynesian and Schumpeterian economists alike would accept this assertion of the key role which technical change plays in fostering economic growth. But, although the causal link between technical change on the one hand and growth on the other is uncontroversial, a large number of often competing theoretical and empirical methods have been used to try to account for this influence of technological innovation on economic growth.

In the late 1950s, the seminal papers by Abramovitz (1956) and Solow (1957) attempted to account for economic growth in the United States, finding it to be not fully explained by the increase in productive inputs such as labour and capital alone. The largest part of growth was thus attributed to a residual which, probably for lack of a more appropriate term, was labelled 'technical change'. Already in 1956, Abramovitz noted that:

since we know little about the causes of productivity increase, the indicated importance of this element may be taken to be some sort of measure of our ignorance about the causes of economic growth in the United States and some sort of indication of where we need to concentrate our attention. (Abramovitz, 1956, p. 133)[5]

In subsequent research, much effort was devoted to trying to understand better the origin of productivity increases by 'squeezing down' the residual (Nelson, 1981), either by introducing other variables such as education, research and development (R&D) and other technology-related factors (on which, see the review by Griliches, 1995), or by adjusting to account for the different quality of inputs (on which see the review by Jorgenson, 1996). Several studies undertook such growth accounting for a variety of countries, allowing international comparisons to be made (see Denison, 1967). These showed clearly that growth rates varied considerably across nations and that differences in technological competence played a significant role. Although this body of literature enlarged the original framework, technology was still treated as a public good.[6]

The basic prediction of the neo-classical theory, based on the notion that the main engine of growth, technology, was a freely available good, was that in the long run all countries should converge towards a similar income level (providing that they were experiencing the same rates of capital accumulation). This hypothesis emerged basically from the actual economic trends of the post-World War Two period with the United States having a large technological advantage over rival countries and transferring know-how and expertise to several allied countries (see Rosenberg, 1972; Nelson and Wright, 1992). This was a crucial factor in helping some of these countries – most notably Japan, Germany and Italy – achieve growth rates higher than the US itself (the leading country). However even during the 1950s and 1960s the assumption of freely available and transferrable technological know-how was unrealistic, with many countries failing to benefit substantially from such know-how produced by the leading country. Growth theory was thus unable to explain why some countries managed to catch up while others fell behind.

5 New departures on growth and technology

Different hypotheses on economic development emerged from a variety of sources. Although the view that a substantial income gap offers an opportunity for catching up, this is no longer credited to free access to the available stock of knowledge. It has been pointed out that emerging countries might avoid repeating the same mistakes and learn at a faster pace than their predecessors (Ames and Rosenberg, 1963). These countries might also benefit from new vintage capital stock and from being able to create a more modern infrastructure than their predecessors at a similar stage of development. However, the broadly institutional theory suggested strongly that such catching up in income is neither automatic nor easy to accomplish. Institutional factors, such as social rigidity (Gerschenkron, 1962), a stratified class structure (Olson, 1982), or an unwillingness to provide incentives for the innovators (Rosenberg and Birdzell, 1986) can seriously hamper the catching up potential of a nation.

These arguments were supported by a large body of evidence, more of a historical than an econometric nature. In the nineteenth century, Germany had managed to catch up in industrial development in part by successfully exploiting and learning from British technology but at the same time Russia had failed, in part due to her failure to have broken with the feudal structure of Russia's countryside. In the second half of the twentieth century, Japan managed to catch up with the United States, while other countries did not. And over the last 25 years in particular, several countries in the East Asian rim managed to take off in industrial development while countries which in the early 1970s appeared to have certain advantages, such as the Latin American nations, failed to so develop. The history of economic development thus shows that growth patterns tend to be related to specific economic, institutional, social and cultural differences across countries.

Generalising from these lessons, it would seem that to close a gap in income, backward nations need to catch up in terms of technological competence. Indeed, there appear to be no examples of a country catching up in terms of technology without also catching up in terms of income. Of course, this leads open the question of causality. Technological competence may be the thermometer of economic development as much as its engine (Dosi, 1982). But backed by an increasing number of empirical studies, the evidence appears to support the view that technological competence is the key to a successful catching-up strategy. A successful strategy for economic development will therefore be associated with the ability of the country in question to create their own endogenous expertise.

Four chapters in this book (Patel and Pavitt, Archibugi and Pianta, Pianta, and Gittleman and Wolff) address the issue of technical change and growth by exploring the conditions which allow nations to catch up in technological competence as well as in income. All of them follow a very different approach from the traditional neo-classical growth theory since they reject the notion of freely available technology. These chapters take as a starting point the fact that the competence levels of nations are highly varied – as indicated by a variety of internationally comparable data on education, labour productivity, the generation of innovations and so on. How important are these technology gaps for predicting countries' growth rates? Within the traditional neo-classical framework, the existence of such gaps is not denied, but they are not considered to be very important in the long run. Non-neo-classical theory, on the contrary, would regard such technology gaps as important determinants of growth rates since the ability to exploit knowledge developed elsewhere will be heavily conditioned by the starting position of each nation. Institutional theory would also hold that the various components of technological competence can play very different roles in development. Education, formal activities devoted to generate innovations, expertise in the capital good sectors, high levels of international integration and so on, are all considered to be potentially important factors in nurturing economic development. But of course, each of these factors can play a different role depending on the nature of the technologies, industries and countries involved.

Gittleman and Wolff (this volume), considering a large number of developed and developing countries, argue that within the restricted club of the industrial market economies, investment in formal R&D activities appears to be related to growth rates but that it does not play any significant role in explaining growth rate differences among developing countries. Education and training play a much more important role in allowing such countries to achieve high growth rates. This result is intuitively plausible: countries at the initial stage of their industrial development might benefit much more by substantial training and education programmes since this will allow them to absorb a vast and still unexploited global stock of knowledge. Investing their own resources to produce original knowledge is certainly beneficial in the long run, but it will require major investments to reach the critical mass needed to secure economic returns. Moreover, formal knowledge needs to be combined with other assets such as an adequate capital stock, a productive infrastructure and so on, which may be lacking in underdeveloped countries.

But after a certain degree of economic development, the strategy of acquiring know-how from abroad without any endogenous effort will become much less effective. Imitative activities become much closer to innovative activities and countries which fail to couple imitation with innovation jeopardise their long-term economic performance. Concentrating on the more homogeneous and integrated group of the OECD economies, Pianta (this volume) shows that the growth paths followed by these nations were nurtured from different sources. For some countries the crucial source of growth has been intangible investment proxied by R&D expenditure, while other countries have concentrated on gross fixed capital formation. Both R&D and investment are crucial engines for development, although each country has a distinctive balance between the two. Significantly, countries with the highest growth rates have combined their efforts in tangible and intangible investments.

Two chapters in this book (Patel and Pavitt, and Archibugi and Pianta) address the issue of convergence from the viewpoint of resources devoted by nations to technological expertise. Focussing on industrial countries only, these studies show that patterns of convergence in the generation of technology (measured by R&D and patent-based indicators) do not emerge strongly, and that in some cases a divergent pattern has occurred over the last 20 years. But the same countries have experienced a limited but significant convergence in GDP per capita. These asymmetric trends between technological and economic indicators need to be explained. It could be argued that the laggard industrial countries within this group managed to catch up in income by acquiring the technology of the leading countries. A number of sources, such as increased trade flows, foreign direct investment, formal cooperation agreements and so on, might have helped this process. The basic question is how far such a convergence in income could go without a similar convergence in technology.

6 Beyond aggregate growth models

The studies of Patel and Pavitt and Archibugi and Pianta also point out that an analysis of technological convergence at the aggregate level is highly unsatisfactory. As outlined in section 3 above, technology is a highly diversified phenomenon. Each country has a specific pattern of technological advantages in some areas and disadvantages in others. Moreover, competence in different technological areas will not have the same impact on growth. Traditional growth theory implicitly assumes that the level of efficiency reached by each country is more important than the choice of its main products: 'what to produce' is less important than 'how to produce'. On the contrary, institutional theory stresses that the sectoral composition of output, as well as of technological competence, is a key factor in determining growth rates. In Freeman's words: 'Economic growth is not merely accompanied by fast-growing new industries and the expansion of such industries; it primarily *depends* on that expansion' (Freeman, 1994, p. 79). Countries which invest their resources in new and expanding industries have, *ceteris paribus*, a better chance of growing than do countries which invest in declining industries.

Already in the early 1960s, the technology gap theory emphasised the crucial importance of sectoral, rather than aggregate, competence differences across countries in explaining economic performance (see Posner, 1961).[7] Thus while the technological lead of a nation in some areas is being eroded by the imitation of other nations, new divergences between the leader and the followers may occur in other rising industries.

Posner and his followers, however, assumed that the world technological capabilities were organised according to a precise hierarchy and that the leading country would have preserved its leadership. Although the gap would have necessarily shifted from one industry to another (from aircraft to semi-conductors, from semi-conductors to biotechnology and so on), it was assumed that the leader would have continued to lead, and the followers to follow. This is not surprising if we consider that this hypothesis was formulated at the beginning of the 1960s when one country, the United States, had a clear lead in most high-technology products.

But this model is no longer a valid characterisation of the contemporary world. On the one hand, the number of countries at the frontier of technological innovation has grown considerably, while, on the other hand, the knowledge base of contemporary society has experienced an impressive expansion. The division of labour between countries now takes place in both new and traditional industries. Patel and Pavitt (this volume), elaborating on while in a sense departing from the notion of technology gaps, identify the advantages and disadvantages of nations across the various technological fields, finding a rather different technological landscape, where each nation has a distinctive and not necessarily hierarchical place in the world's division of labour. While Japan is still lagging behind in nuclear technology *vis-à-vis* the United States, the United States is now lagging behind Japan in photography and several other industries. This has been systematically measured by looking at the sectoral strengths and weaknesses in scientific and technological activities (see Archibugi and Pianta,

1992), and by looking at the technological strengths of national large firms in a number of patent classes (see Patel and Pavitt, this volume).

It is certainly significant that the differences in the technological profiles of countries are increasing, as shown by Archibugi and Pianta in this book. This has important implications for the debate on convergence and catching up. Although a partial convergence in efforts to generate and assimilate innovations is occurring among a restricted club of countries, this does not mean that countries are converging towards a similar profile of sectoral specialisation. It rather seems that convergence in the aggregate amount of resources devoted to innovation is coupled with an increased division of labour in the generation of new technologies.

7 High technology and specialising for growth

Moving away from standard aggregate macroeconomic models, a sectoral analysis of nations' innovative activities has some important implications for growth theories. As already argued, the growth rate achieved by a country is not only related to the aggregate amount of resources devoted to innovation. This is surely a fundamental part of the story, but it is equally important to look at the sectoral composition of these efforts. It can no doubt be readily accepted that countries which concentrate their resources in the new and growing technological fields will have, other conditions being equal, a higher probability of growing faster; the question is, which are these fields?

It is generally assumed that the so-called 'high-tech' industries better serve nations' competitive performance. World trade in high-tech industries, and especially in the high-tech electronic products, has grown much more than in other manufacturing industries (see Guerrieri and Milana, this volume). This evidence corroborates the general view that high technology means high growth. However, the definition and meaning of 'high-tech' industries is not unproblematic. Grupp and Münt (this volume) argue that the concept of high technology is not well defined in economics, proposing a more accurate distinction between 'leading-edge' and 'high-level' technologies. Guerrieri and Milana (this volume) challenge the standard measures based on the R&D intensity of industries and propose a more complex approach based on expert judgements.

Despite some pioneering studies,[8] however, we know little about the role played by crucial band-wagon sectors in industrial development. Economic historians have shown the importance of emerging industries for development; we generally associate the industrial revolution with steam engines, textile machinery and railways and the public opinion of our age rightly associates the contemporary economic transformations with computers, software and telecommunications. But economic theory is still lagging behind: while interesting industry case studies are available, and sophisticated multisectoral growth models have been developed, a systematic exploration of the role played by rising sectors in economic growth is still lacking. It is to be hoped that research on innovation and development will fill this gap over the next few years.

8 Innovation and international competitiveness

It has long been a subject of debate whether, in the classical example of comparative advantage (Portuguese wine and English cloth) David Ricardo (1911) was implying more importance to nations' resource endowments or to nations' technological competence. Whatever Ricardo's thoughts on the matter, for too long international economics paid too much attention to resource endowments and too little to technological competence (see Krugman, 1995). In the last decade, the situation has changed, as indicated by the growing technology gap trade theory and, more recently, by the new growth theory.

It is now generally accepted that advantages in technological competence will lead to a better performance in foreign trade. There are at least three links between innovation and international competitiveness. First, process innovations reduce production costs and hence output prices, increasing competitiveness. Second, minor product innovations improve the quality of commodities and make them more appealing in both domestic and foreign markets. Third, major product innovations create, for a limited period of time, a monopolistic position which helps to impose those products in the market, while at the same time bringing in monopoly profits.

As argued above, the technology gap trade theory has since the early 1960s (Posner, 1961) stressed the crucial role of innovation as a determinant of export market shares. In this book, this hypothesis is tested by Amendola, Guerrieri and Padoan, exploiting a detailed database which combines trade and patent data for ten industrial countries and 38 industries. The authors show that the differences between countries in terms of trade specialisation are larger than in terms of technological specialisation, reinforcing the view that advanced industrial economies required a wide technological base.

Research also shows that the division of labour among countries, at least in manufacturing, evolves very slowly, corroborating the hypothesis of a cumulative pattern of technological competence. From a nation's perspective, it is certainly much more difficult than predicted by Posner to move from an established competitive advantage in one industry to another. As suggested by Nelson and Winter (1977) and Pavitt (1988a), expertise moves along 'natural' trajectories of technological accumulation. A country leading in the chemical industry will probably have an advantage over other nations in entering new and related fields such as new synthetic materials, but it is much more difficult for a leading country in aircraft or nuclear technologies to exploit these advantages in, say, consumer electronics. As far as technological competence is concerned, sector-specific accumulation prevails over leap-frogging.[9]

The third and most significant result of Amendola, Guerrieri and Padoan's research is that technological innovation is a crucial factor in explaining trade patterns at the sectoral level. Countries manage to have a good performance in international trade in the sectors where they also excel in terms of technological innovation. The comprehensive time period covered provides a solid test of the role of technology in international competitiveness. It is however important to note that this excercise moves away from the traditional technology gap approach since it no longer assumes the existence of an established hierarchy among nations. Looking at the comparative, rather than absolute advantages, the authors show that the 11 nations considered in the test have a mixed pattern of specialisation in both new and traditional technologies. It is certainly significant that the United States has trade and technology advantages in resource-based as well as technology-intensive industries.

What are the implications of this empirical evidence? One implication should be to allow us to understand the advantages to be had for any one country from innovating to a greater extent than its competitors, and over a wider range of industries. In the short term, these benefits will translate into a surplus in the trade balance. In the longer term, innovative nations will have two main advantages: firstly, improved terms of trade and secondly the ability to specialise in whatever proves to be the most rewarding industries. Both of these could prove crucial factors in allowing a nation to achieve higher growth rates.

Grupp and Münt (this volume) also show that the sources of competitive trade advantage are substantially different for the two groups of leading-edge and high-level technologies. While leading-edge technologies rely mainly on scientific inputs and involve heavy interactions between public and business R&D, high-level technologies are much more dependent on firms' decisions to invest in innovation. The empirical test for the European countries shows that a strong division of labour has occurred within the region, with Germany and the Netherlands focussing on high-level technologies and France and the United Kingdom on leading-edge technologies.

It is well known that globalisation implies much more than international trade. Firms can use a variety of methods to exploit their innovations, which include foreign direct investment, non-equity collaborations, granting of licences and so on. We are well aware of these aspects although they are not addressed directly in this volume (but on which see the various contributions in Archibugi and Michie, eds., 1997). New and more complex forms of internationalisation, however, have not reduced the crucial importance of trade. International competitiveness shaped by innovation remains one of the fundamental determinants of nations' comparative performance and the pursuit of such competitiveness has been crucial to public policy in the successful countries.

9 The importance of the domestic market

Given that countries' specialisation in trade and technology matters for their growth rates, what are the determinants of a country's position within the international division of labour? Following the suggestions of Kaldor and Linder, Fagerberg (this volume) argues that the domestic market can play a crucial role in fostering a nation's ability to excel in certain products rather than in others. The role of 'advanced domestic users' has been emphasised by Porter (1990), while Pavitt (1984) and von Hippel (1988) have argued convincingly that users can play a crucial role in promoting innovation, and this often takes place in systems of innovation which are national in scope (Lundvall, 1993; Nelson, ed., 1993; Freeman, 1995).

The importance of the domestic market is tested by Fagerberg for a sample of 16 advanced countries and 23 different product groups in three different years (1965, 1973 and 1987). The sample considers the export performance of mainly intermediate products and links them to the expected main users in the domestic market (see Fagerberg, this volume, table 10.1). This implies that the linkages explored are of a vertical nature (such as textile machinery–textiles, tractors–agricultural products and so on). Fagerberg's results confirm that the domestic market plays an important role in fostering international competitiveness. He also shows that nations open to foreign trade exploit their own domestic user–producer linkages better. Given that the domestic market is an important stimulus to achieving international competitiveness, targeted industrial policies can thereby boost a country's trade position and hence growth rates.

10 New departures

The studies published in this volume invite fresh exploration of the connections between three epoch-making phenomena: technical change, growth and trade. The generation and impact of innovations, the factors which rule economic growth and their differences between countries, and the role of trade in the organisation of the world economy, are all issues deserving of further analysis. Such work can best be done by exploring the interdependence between these phenomena. Three main recommendations for further research seem to emerge from the following chapters.

Complexity

Technology is a complex phenomenon. Although the importance of technology has become widely recognised by mainstream economics, as indicated by the new growth and new trade theories, little attention has so far been devoted to accounting for its complexity. On the other hand, the contributors to this book have chosen to formulate empirically testable questions rather than abstract models and the results, even when rough, tell us something about the real nature of innovation which is an indispensible component for the construction of a theory of economic growth.

Level of aggregation

The main patterns and regularities which emerge from complexity can be discerned with the help of detailed case studies and quantitative economic studies; the chapters in this book have focussed particularly on the latter, undertaking especially cross-industry and/or cross-country research. We have already pointed out that there is a significant difference in the two literatures on innovation/growth and innovation/trade. Much of the innovation/growth literature is at the aggregate level only, ignoring the crucial importance played by sectoral differences. Much of the innovation/trade literature is on the contrary at the industry level, although it often overlooks the macroeconomic implications. We believe that some 'spillovers' might be beneficial to both these bodies of literature. The influence of technology on growth will become much clearer once it is accepted that not all knowledge has the same impact on economic performance. Likewise, the studies on technological competence and trade should devote more attention to the macroeconomic and long-term implications.

Policy

Growth and competitiveness are two areas of major policy concern. The advancement of knowledge is also an area of active policy action, particularly since it has become clear that it has important feed-backs on economic performance. Too often, however, economic research in this field fails to focus on any clear policy questions, hampering the impact of such studies. The economics of innovation, trade and growth can only benefit from a clearer focus on policy options and questions. This requires theorising grounded in economic, social and institutional realities which will not only increase the usefulness of such work but may also sharpen the focus. The work reported in this book represents, we believe, an important step in this direction, with theoretical discussion grounded in just such an approach, a wealth of new empirical research, and clear policy discussion pointing not only to the importance of achieving an innovative economy but pointing also to the sort of public policy initiatives which would help to achieve such goals.

Notes

1 On the separate issue of the direct relationship between growth and trade, which is not addressed directly in this book, see Kitson and Michie (1995a, 1995b).

2 For surveys of the economic theory of technology policy, see Metcalfe (1995a, 1995b), Mowery (1995) and Dasgupta and Stoneman (1987).

3 For related research on the themes of technical change, growth, and trade following broadly the same approach taken in this volume see especially Dosi, Pavitt and Soete (1990), Silverberg and Soete, eds. (1994), and Fagerberg, Verspagen and von Tunzelmann, eds. (1994).

4 Although Arrow (1962) is responsible for this mainstream formulation, he has been much more open minded in questioning this assumption than many of his followers. See, for example, Arrow (1994). New growth theory has dropped the hypothesis that knowledge is a freely available good. See Romer (1990) and Grossman and Helpman (1991). But other recent growth literature still regards technology as freely accessible, see Mankiw (1995). For a critical review of this literature, see Verspagen (1993b).

5 That this statement was an accurate representation of the point of departure in the field is indicated by the fact that it is probably the single most cited phrase in the economics of growth.

6 For a comprehensive review of the literature on growth rate differentials, see Fagerberg (1994).

7 The importance of an analysis at the sectoral level emerged especially in the technology-gap trade theory (see Soete, 1981, 1987) and it is reaffirmed in the chapters by Amendola, Guerrieri and Padoan, and Guerrieri and Milana in this book.

8 Meliciani and Simonetti (1996) and Breschi and Mancusi (1996) develop a methodology to identify the fast-growing and the declining technological classes, and have used this to indicate the comparative position of individual nations.

9 Cantwell (1989) has shown that patterns of technological accumulation tend to be stable across long historical periods.

2 The economics of technical change

CHRISTOPHER FREEMAN

1 Introduction

One of the continuing paradoxes in economic theory has been the contrast between the general consensus that technical change is the most important source of dynamism in capitalist economies and its relative neglect in most mainstream literature. Those economists, such as Marx in the nineteenth century and Schumpeter in the twentieth, who attempted to assign a more central role to technical innovation, were regarded as rogue elephants, whose work, although certainly of interest, should not be taken too seriously.

This paradox was repeatedly pointed out in survey articles (e.g., Kennedy and Thirlwall, 1973), in books on invention and innovation (e.g., Jewkes, Sawers and Stillerman, 1958) and in histories of economic thought (e.g., Blaug, 1978). Various explanations were advanced, the most frequent being the 'black box' explanation – that technical change was outside the specialised competence of most economists and had to be tackled by engineers and scientists. This approach fitted quite well with the convenient (but erroneous) assumption that science and technology could be treated as exogenous 'manna from heaven' and need not be examined in any depth for most purposes.

Jewkes *et al.* (1958) advanced two other reasons for the relative neglect of technical change by most economists: the lack of quantitative data and the preoccupation of many economists in the 1930s and 1940s with employment and business cycle problems. Both of these explanations are revealing of certain attitudes of mind. If measurement difficulties inhibited work on the most important issues, the response in most disciplines would be to concentrate effort on tackling the measurement problems, not to concentrate attention on other issues.

Furthermore, the 'Keynesian' problems with which many economists were preoccupied in the 1930s (and again today) were in any case not

16

unaffected by technical change and could not therefore be considered as a separate alternative agenda. Yet although Jewkes and his colleagues made a major contribution to the economics of invention (which we shall discuss later), they clearly did regard this as something separate from the rest of theoretical and applied economics – as a specialisation which did not affect other work.

This attitude is still prevalent, both among those who do specialise in the economics of technical change and those who do not, but it has been very much eroded in the 1980s and 1990s. There has been a marked upsurge of interest which is evident both in mainstream economics and an explosion of new specialised journals (for example, *Journal of Evolutionary Economics, Economics of Innovation and New Technology, Industrial and Corporate Change* and *Structural Change and Economic Dynamics* all commenced publication in 1990–1). In the 1950s Everett Rogers (1961) could find only one empirical study of the diffusion of innovations by an economist but by 1986 he found dozens and it is much the same story in relation to firm behaviour, growth and development economics, international trade and several other areas. There is now a far greater readiness to look inside the 'black box' (Rosenberg, 1982) and study the actual processes of invention, innovation and diffusion within and between firms, industries and countries.

In deference to the work of Schumpeter in the first half of the century, it has been customary to refer to much of this work as 'neo-Schumpeterian' or 'evolutionary' and virtually all those involved share the fundamental postulate of Schumpeter (and Marx) that capitalism is an economic system characterised above all by evolutionary turmoil associated with technical and organisational innovations. Schumpeter never liked the idea of 'disciples' and advised his readers to regard his work as only a first approximation which should not be a dogma but a set of ideas to be revised and amplified in the light of new evidence. This advice was followed with greater alacrity by the neo-Schumpeterians than somewhat similar advice given by Marx. Although a 'Schumpeter Society' was established in the 1980s and shows every sign of flourishing, it is a broad church and most neo-Schumpeterians have not hesitated to criticise some of Schumpeter's main propositions, including as we shall see, his basic concepts of innovation, diffusion and entrepreneurship (see for example, the proceedings of the biennial conferences of the Schumpeter Society: *Evolutionary Economics* (Hanusch, ed., 1988), *Studies in Schumpeterian Economics* (Heertje and Perlman, eds., 1990) and *Studies in the Schumpeterian Tradition* (Scherer and Perlman, eds., 1992), all of which illustrate this diversity of approach). The neo-Schumpeterians have criticised Schumpeter's work very much in the spirit of his own advice, i.e., on the basis of new empirical research evidence. They have also tackled topics which he almost completely neglected, such as underdevelopment, international trade and regional development. Consequently, the description 'neo-Schumpeterian' is used here in a very broad sense to indicate the scope of the subject matter, rather than an ideological standpoint. It includes work which could in many respects be described as neo-classical as well as much which could certainly not be so described.

This article will concentrate on those topics where the results of empirical research have been most impressive and where they have presented the greatest challenge to established theory. The primary focus will be on innovations and their diffusion at firm and industry level. This means that many other topics cannot be satisfactorily surveyed. Thus, for example, the issue of technical change and employment is entirely neglected and so too are environmental issues, energy, business cycles and the role of military technology. Even though firm-level technical activities are discussed, the more specialised area of project selection and evaluation techniques is not. Some subjects which are of the greatest importance are hardly touched upon, such as international trade and growth theory. These are areas where neo-Schumpeterian research has made a considerable contribution, but fortunately there are major recent reviews of the literature in Dosi, Pavitt and Soete (1990) for international trade and Verspagen (1992) for growth theory. Nor does this review take up themes which have been largely neglected by neo-Schumpeterians, such as consumer behaviour, patterns of consumer demand (with outstanding exceptions such as Pasinetti, 1981) or financial innovations (again, with outstanding exceptions, such as Heertje, ed., 1988; Mowery, 1992a; Christensen, 1992). Finally, the survey does not cover government policies for technology and innovation, which is again an area where neo-Schumpeterian research has made a major contribution (see, for example, Nelson et al., 1967; Nelson, 1984; Pavitt and Walker, 1976; Tisdell, 1981; Rothwell and Zegveld, 1982; Teubal, 1987; Krauch, 1970, 1990; Meyer-Krahmer and Soligny, 1989; Ergas, 1987; Edquist, 1989; Stoneman, 1987; Salomon, 1985; Dasgupta and Stoneman, eds., 1987; Sharp and Holmes, eds., 1988; Soete, 1991b; Hilpert, ed., 1991; Smith, 1991; Limpens, Verspagen and Beelen, 1992; Arundel et al. 1993). Fortunately, here too, there are comprehensive recent reviews (OECD, 1991a). A major publication (Stoneman, ed., 1995) will review many of those topics which are not dealt with here, including government policies, financial innovations, international trade, technical change and employment.

There is another reason for the main part of this chapter to concentrate on the results of micro-level studies of innovations and their diffusion at the firm and industry level. In doing so, it takes into account the major critical comments made in earlier literature surveys in the 1970s – those respec-

tively by Kennedy and Thirlwall (1973) and by Nelson and Winter (1977).
After a thorough examination of the relevant literature of the 1950s and 1960s (they cited nearly 300 references), Kennedy and Thirlwall concluded:

We have tried to survey as comprehensively as possible, within the space permitted, the major lines of applied research pursued by economists on the subject of technical progress. The main conclusion can be summed up very briefly. Macro-studies of technical progress which seek to estimate the rate of technical progress as a residual component of the growth of output are fraught with aggregation and identification difficulties but almost invariably find technical progress as the prime determinant of the rate of growth regardless of the production function specified. Technical progress and advances in knowledge are not synonymous without some adjustment for increases in output due to movements towards *known* production possibilities, but even studies which make some adjustment frequently make 'pure' technical progress the single most important determinant of the growth of living standards. For this reason alone the process of technical change is of vital interest. The reader of this survey may well have been struck by the apparent thinness of studies in this field as compared to macro-economic production function studies. While it is possible we may have missed some important 'micro' studies of technical change, we are nevertheless convinced that the general impression given is a fair one. (p. 166)

In their forward looking paper 'In search of a useful theory of innovation', Nelson and Winter also concluded that micro-level and industry-level studies of technical change were likely to yield more fruitful results than the seductive but blind alley of aggregate production function models.

We would argue, however, that the breadth and strength of the production function framework is inherently limited. To obtain a more solid understanding of innovation and what can be done to influence innovation, it is necessary to study in some considerable detail the processes involved and the way in which institutions support and mold these processes. Since the 'production function framework' contains at best a rudimentary characterisation of process and relevant institutional structure, a considerably more fine grained theoretical structure is needed for these more microscopic studies. (p. 46)

Nelson and Winter argued in particular that variation between industries as diverse as agriculture and aircraft necessitated industry-specific analysis, including differences in the selection environment as well as technological trajectories. Their emphasis on trajectories arose from their challenge to neo-Schumpeterians: how can ordered patterns of innovation emerge despite the industrial diversity and the uncertainty inevitably associated with innovation? How can structure, order and patterns emerge from apparently chaotic variety? Section 7 of this review will deal with some attempts to address this question.

The plea of Nelson and Winter for attention to industrial diversity, for more micro-level studies, and for deeper understanding of trajectories has evoked a significant response as recognised by Dosi (1988) in his more recent survey in the *Journal of Economic Literature*. It is now possible to make far more reliable generalisations about innovations and their diffusion, both in specific industries and for the entire economy than was possible 50 or 60 years ago. Schumpeter had very little to go on in the way of case studies and he himself did not attempt to carry out empirical studies of innovation (Heertje, 1977; Svedberg, 1991; Andersen, 1992b, 1993; Angello, 1990; Shionoya, 1986). It is hardly surprising that the results of the research which has been performed since his death considerably modify his pioneering formulations. Nevertheless, his work is an essential point of departure. Before addressing the main findings of neo-Schumpeterian research on technical change in sections 3–7, this survey therefore first summarises the extent of the departure from Schumpeter's own theoretical framework. Section 8 presents some brief conclusions.

2 The departure from Schumpeter

Schumpeter (1939) advised his followers to study business histories, company reports, technical journals and histories of technology in order to understand the behaviour of the economic system. He even went so far as to maintain that economics should only be a postgraduate subject and that undergraduates should study history and mathematics before tackling economics. He particularly admired the achievements of the great American entrepreneurs at the close of the nineteenth century and the early twentieth century and, as recent research has shown, was strongly influenced by the ideas of Nietzsche and other proponents of 'super-man' theories (Andersen, 1993; Svedberg, 1991).

In contrast to the concept of 'representative agents' enjoying equal access to reliable information and capable of rational calculations about the rate of return on future investments, Schumpeter postulated two types of agent: exceptional individuals (entrepreneurs) who, although certainly not able to foresee the future, were willing to face all the hazards and difficulties of innovation as 'an act of will', and a second much more numerous group of 'imitators', who were merely routine managers and followed in the wake of the heroic pioneers in the first group.

Whilst preserving Schumpeter's emphasis on the role of uncertainty and on the importance of innovation, neo-Schumpeterian research has moved a long way from this somewhat romantic model. Schumpeter (1928, 1942) himself moved some way from his own original (1912 [1934]) formulation. In fact, he moved so far that Phillips (1971) distinguished a 'young' Schumpeterian model from an 'old' one (Freeman, Clark and Soete, 1982).

In his later years, Schumpeter recognised that in large firms innovation had become bureaucratised and that organised and specialised R&D departments played an increasingly important role in the innovative process. Indeed, he went so far as to maintain that a development engineer in the R&D department of a large electrical firm could be an 'entrepreneur' in his sense of the word (Schumpeter, 1939).

This led him also to stress the predominant role of large oligopolistic firms in technical innovation and to a more charitable view of monopoly than that prevalent in orthodox theory and policy making. After his death many attempts were made to discredit this part of his work, which was often (wrongly) assumed to be his main contribution to economic analysis and known as the Schumpeterian theorem. The controversy on innovation, size of firm and market structure rumbled on for decades (see Kamien and Schwarz, 1975, 1982; Scherer, 1965, 1973, 1992; Sylos Labini, 1962; Soete, 1979) and, as we shall see, was one of those long-continuing debates partly resolved by new empirical research in the 1970s and 1980s (section 5). Another such long-running theoretical controversy was that surrounding 'demand-pull' and 'technology-push' theories of technical change. Schumpeter was a clear proponent of entrepreneurial technology push. Here too we shall attempt to show that the results of micro-level empirical research have largely resolved this ancient debate (section 6). Interactive systems models have replaced simple linear models whether led by markets or by technology.

The results of the empirical research reviewed in this survey confirm the main assumptions of 'bounded rationality' postulated by Simon (1955, 1959, 1978, 1979), Simon et al. (1992) and other behavioural economists studying decision making under conditions of uncertainty (e.g., Cyert and March, 1963; March and Simon, 1958; Heiner, 1983, 1988 and see also Dosi and Egidi, 1991). The weight of this evidence is such that those who wish to rescue the extreme rationality–optimisation paradigm have been obliged to concede that above all in relation to innovation, the 'classic defence' of this paradigm is 'not descriptive of the actual process by which decisions are made' (Winter, 1986a). They have had to retreat into 'as if' propositions, and to an evolutionary model: firms do not actually take optimising decisions based on accurate information or rational expectations but those that survive in competition have supposedly behaved as if they did (Alchian, 1950; Friedman, 1953; Lucas, 1986).

However, this defence is no more tenable than its predecessors. As Hodgson (1992, 1993) has shown, biologists themselves do not claim 'optimisation' through natural selection. Darwin (1859) pointed out that the improvements brought about by natural selection were relative. 'Natural selection will not produce absolute perfection' (p. 202). In any case, biological evolutionary models provide unsatisfactory analogies for social scientists, since human behaviour has many features which are unique (Penrose, 1952; Clark, 1990; Saviotti and Metcalfe, eds., 1991; Freeman, 1991b; Hodgson, 1991, 1992) and social systems do not evolve on Darwinian principles of natural selection. The 'selection environment' includes many features, which are absent in biological evolution and the purposive design activities of human beings, whether in technology or in institutions, have no parallel elsewhere in the animal world.

Just as Keynes showed that the economy could be locked into a less-than-full employment situation, the empirical evidence is again strong that technological systems can be 'locked in[to]' sub-optimal solutions through a succession of small events. Paul David (1976, 1985, 1986a, 1986b, 1992, 1993), David and Greenstein (1990), David and Steinmuller (1990), Brian Arthur (1983, 1986, 1988, 1989) and his colleagues at the Santa Fe Institute (1990–3) have been particularly resourceful in demonstrating the causes and consequences of 'lock-in', whether by standardisation, compatibility standards or other reinforcing institutions, for evolution of the economy and for model building. As some leading neo-classical theorists (e.g., Hahn, 1987) have themselves acknowledged, history always matters. Path dependence and irreversibility rule, not hyper-rationality. The implications for general equilibrium theory are of course considerable, although many who work in this tradition and even in the neo-Schumpeterian tradition appear reluctant to recognise them. In this of course, they also follow Schumpeter who never quite cut the Walrasian umbilical cord. Esban Andersen (1992b, 1993) has shed interesting new light on this aspect of Schumpeter's work.

The empirical research surveyed in sections 3–7 also demonstrates the limitations of Schumpeter's theories of innovation and entrepreneurship. Ruttan (1959) put the point a bit too strongly when he said that: 'Neither in *Business Cycles* nor in Schumpeter's other work is there anything that can be identified as a theory of innovation.' Nevertheless it remains true that Schumpeter's emphasis on the role of entrepreneurship as an act of will and his tendency to belittle subsequent adopters of innovations as 'mere' imitators did tend to obscure many important aspects of innovation and diffusion, which have been illuminated by later research (section 6).

This research points strongly to the cumulative aspects of technology, the great importance of incremental as well as radical innovations, the multiple inputs to innovation from diverse sources within and outside the firm and the changes made to innovations by numerous adopters during diffusion, both within and between countries. It is true that the empirical research does often confirm the importance of individuals variously described as 'product champions' (Schon, 1973), 'business innovators' (Freeman, 1974), or 'network coordinators' but they are sometimes hard to identify within a more anonymous process in which pygmies play an essential part as well as giants (see also Roberts, 1991). We now consider in more depth this complex process of innovation in firms (sections 3 and 4) and industries (section 5).

Section 3 summarises the work which has been done on the multiple external linkages of the firms, both with sources of scientific and technical information and knowledge, and with other firms, both users of its products and systems and sub-contract suppliers. The research shows that the view of the firm as a passive recipient of 'information' is untenable. Knowledge accumulation is an interactive process and the flows of information and knowledge are just as important for understanding firm behaviour and that of the economy as the flows of materials, components and intermediates.

Section 4 deals with the learning process and knowledge accumulation within the firm and introduces the considerable discussion of the special characteristics of the Japanese firm. The Japanese techniques for managing innovation appear to have been specially successful in the continuous improvement of products and processes. The neo-Schumpeterian research on incremental innovations and radical innovations is taken up in section 5, together with the complex problems of definition and measurement. This section also summarises the response of neo-Schumpeterians to the challenge of Nelson and Winter on industry-specific studies and the problems of developing a taxonomy of industries.

3 Technical learning from external sources

The strong emphasis in much neo-Schumpeterian research on firm-specific technological knowledge accumulation (e.g., Teece, 1988; Teece et al., 1990; Foray, 1987; Amendola and Gaffard, 1988; Pavitt, 1986a, 1986b, 1988a, 1990; Teubal, 1987; Gaffard, 1991; Granstrand, 1982; Granstrand and Sjölander, 1992; Eliasson, 1990, 1992; Dosi, 1984; Stiglitz, 1987; Swann, ed., 1992; Achilladelis, Schwarzkopf and Lines, 1987, 1990; Miller et al., 1993; Winter, 1987) should not be taken to mean that exogenously generated scientific discoveries and advances play no part in technical innovation at firm level. On the contrary, much of the recent empirical work, like the earlier studies of Carter and Williams (1957, 1958, 1959a, 1959b) points to the importance of contacts with the world of science and to the increasing interdependence of science and technology (Nelson, 1962; Freeman, 1974; Mansfield, 1980; Price, 1984; Grupp, ed., 1992; Rosenberg, 1982, 1990, 1992, 1994).

A particularly important point made by Pavitt (1993) in his paper on 'What do firms learn from basic research?' is that the contribution made by basic science to industry is mainly indirect, in the form of young recruits with new and valuable skills and knowledge, rather than direct, in the form of published papers (though these too can of course be very useful). The Yale University Survey of 650 US industrial research executives (Levin *et al.*, 1987) showed that basic scientific skills and techniques in most disciplines were valued more highly and rated as more relevant than academic research results.

However, although less frequent and directly affecting a much smaller number of firms and innovations, inputs of new scientific knowledge are nevertheless extremely important (Nelson, 1959a, 1959b, 1962). The results of the studies on radical innovations commissioned by the US National Science Foundation (NSF) (Project TRACES, 1969) demonstrated not only that the major twentieth century innovations would have been impossible without the prior accumulation of scientific knowledge, but also that some very recent scientific advances played a critical part during the development stage (see also Mansfield, 1991).

Later studies of innovation have amply confirmed these results (e.g., Pavitt, 1971). Project SAPPHO (Freeman, 1974; Rothwell *et al.*, 1974) showed that in the chemical and instrument industries, the ability to make use of external sources of scientific expertise and advice was one of the main determinants of success. The Manchester study of Queen's Award winning innovations (Langrish, Gibbons, Evans and Jevons, 1972) also confirmed the NSF conclusions and showed in particular the great importance of informal contacts with University scientists in several industries (Gibbons and Johnston, 1974). It is important to note that all of these studies and especially the Yale survey demonstrated that the nature, depth and frequency of this interaction was highly industry specific (see section 5) and varied also with the nature of the innovation. For incremental innovations, scientific skills may also still be very important, but results of recent academic research will rarely be significant for them except in a few technologies, where the science is almost indistinguishable from the technology.

More recent research on success and failure in innovation (see Rothwell, 1992; Van de Ven *et al.*, 1989) has generally confirmed these conclusions whilst also demonstrating the role of corporate strategy and government policy in developing networking relationships with external sources of information, knowledge and advice (Dodgson,1991, 1993; Teubal *et al.*, 1991; Sharp, 1985; 1991; Coombs and Richards, 1991; Coombs, Saviotti and Walsh, 1992; Carlsson and Jacobsson, 1993; Steele, 1991). The new generic technologies which diffused rapidly in the 1970s and 1980s (information and communication technology (ICT), bio-technology and new materials technology) have been shown in numerous studies to intensify the science–technology interface and to enhance the importance of external networks for innovative success (see, e.g., Orsenigo, 1989, 1993; Dodgson, ed., 1989, 1991; Faulkner, 1986 for bio-technology; Lastres, 1992 and Cohendet *et al.*, 1987 for new materials technology; Nelson, 1962; Gazis, 1979; Dosi, 1984; Molina, 1990; Antonelli, 1986, 1992, 1993; Antonelli, ed., 1992; Lundgren, 1992; and Freeman, 1991a for information and communication technology). The intensity of the interaction between science and technology has also been demonstrated in the 'scientometric' literature using citation analysis and similar techniques, notably by Narin and his colleagues (Narin and Noma, 1985; Narin and Olivastro, 1992) and in the work of the Leiden Science Studies Unit (Van Vianen, Moed and Van Raan, 1990).

Much empirical research (e.g., Lundvall 1985, 1988; Lundvall, ed., 1992) has also shown that another major determinant of innovative success lies in the nature and intensity of the interaction with contemporary and future users of an innovation. In the case of incremental innovations, especially, but also for radical innovations, this has often been shown to be a decisive factor (see section 5). It was one of the main findings of the SAPPHO project and the Manchester project already referred to. Von Hippel (1978, 1980, 1988) and Slaughter (1993) have shown that users may often take the lead in stimulating and organising innovation. So important is the interaction with users that it has become one of the key topics in the research on 'national systems of innovation' (Lundvall, 1993; Lundvall, ed., 1992; Andersen, 1991, 1992a; Fagerberg, 1992; Mjøset, 1992; Nelson, 1990a, 1992a, 1993) and globalisation of technology (see section 7).

The picture which thus emerges from numerous studies of innovation in firms is one of continuous interactive learning (Stiglitz, 1987; Lundvall, ed., 1992). Firms learn both from their own experience of design, development, production and marketing (section 4) and from a wide variety of external sources at home and abroad – their customers, their suppliers, their contractors (a particularly important aspect of Japanese firm behaviour, see Imai, 1989; Sako, 1992; Dodgson and Sako, 1993) and from many other organisations – universities, government laboratories and agencies, consultants, licensors, licensees and others. They also learn from their competitors through informal contacts and reverse engineering. The precise pattern of external and internal learning networks varies with size of firm, but all firms make use of external sources (Foray, 1991, 1993; Kleinknecht and Reijnen, 1992a).

The characteristics of the network also vary with the type of technology and innovation (process, product, service, organisation, incremental, radical), with the sector of industry and with the national environment or 'system of innovation' (see section 7). Equally varied are the methods of learning. For example, learning from other firms may be through informal contacts and 'informal trade' in knowledge (von Hippel, 1982, 1987), through formal collaboration and joint ventures of various kinds (Hagedoorn, 1990; Hagedoorn and Schakenraad, 1990, 1992), through licensing and know-how agreements (OECD, 1988), through recruitment of people, through acquisitions, through reverse engineering and, of course, through espionage (see Mansfield, 1985 on 'how quickly does technology leak out?'). The Yale Survey (Levin et al., 1987) showed that the great majority of US industries regard their own R&D and reverse engineering as the most effective forms of learning, but this survey may have underestimated the role of informal contacts and of espionage. It is also possible that a survey addressed to industrial R&D executives would contain some bias in the replies towards in-house R&D. But clearly a combination of various forms of external and internal sources of learning is essential.

The work of Häkanson and Johansson (1988) and Häkanson (1989) showed that informal flows of information between users and suppliers were actually more important than formal arrangements. However, Hagedoorn and Schakenraad (1990) and several other studies (e.g., OECD, 1986; Chesnais, 1988a, 1988b; Mowery, 1988, 1989) have surveyed the formal arrangements for technology cooperation and exchange and have demonstrated a rapid increase in the 1980s. It should also be remembered that many firms make use of several different arrangements simultaneously. Acs (1990) showed that even within the small science-based firms in Maryland almost all make use of more than one different form of technology collaboration, while the largest firms may have more than a hundred different agreements affecting different products and technologies. Kleinknecht and Reijnen (1992a) showed in an extensive survey that small firms are involved in technical collaboration to at least the same extent as large firms (see also Smith et al., 1991; Rothwell, 1991).

Although external sources of technology and cooperation with other organisations are important in all sectors of industry and for all sizes of firms (Kleinknecht and Reijnen, 1992), an important result of the work of Hagedoorn and Schakenraad (1990, 1992) was to show that the three major generic technologies had led to an explosion of new cooperative arrangements in the 1980s with information and communication technology accounting for the lion's share.

Building on the work of Coase (1937, 1988) and Williamson (1975, 1985), a number of neo-Schumpeterian studies (e.g., Goto, 1982; Fransman, 1990; Foray, 1991, 1993; Imai, 1989; Sako, 1992; Teubal et al., 1991; Imai and Baba, 1989) have suggested that the growth of networking arrangements means that both market and hierarchical relationships are being superseded by new forms of organisation. Others have suggested that this may be associated with the diffusion of information and communication technology, providing both the necessity for collaboration and better means for achieving it. Still others (e.g., Bressand and Kalypso, eds., 1989; Bressand, 1990) have pointed to the tendency for today's networks to become tomorrow's monopolies. A good deal of the empirical evidence points to the view that networking can be explained more in terms of strategic behaviour of firms than in terms of costs, whether transaction costs or others (e.g., Hagedoorn and Schakenraad, 1990). Here is a very fertile field for further illumination of as well as future research into (whether by neo-Schumpeterians or others), with major policy implications for (DeBresson, 1989, 1993) the market structure/innovation debate. Whilst a number of neo-Schumpeterian studies have justifiably emphasised the role of *trust* in networking relationships (e.g., Sako, 1992; Lundvall, 1993; Sabel, 1993) the role of *power* and fear is often neglected.

Much of the discussion on networks relates to Japanese firms (Sako, 1992; Imai and Baba, 1989; Tanaka, 1991; Imai, 1989; Goto, 1982; Schonberger, 1982; Kodama, 1990, 1991; Friedman and Samuels, 1992; Fransman, 1990) and the internal organisation of Japanese firms is often supposed to differ also from typical American and European firms. This question is discussed in the following section and again in the final discussion in sections 7 and 8 in relation to national differences in systems of innovations.

4 Technological learning from internal sources

The SAPPHO project (Freeman, 1974; Rothwell *et al.*,1974) had already shown that good internal coupling between design, development, production and marketing functions was one of the decisive conditions for successful innovation. Many failures could be attributed to lack of communication between the R&D, production and marketing functions, as was also shown in the brilliant sociological study of Burns and Stalker (1961).

Whereas they found cases where R&D personnel had never set foot in a factory, in his comparison of Japanese and American electronic firms, Yasunori Baba (1985) defined Japanese development strategies as 'using the factory as a laboratory'. Takeuchi and Nonaka (1986) have described Japanese firms as playing rugby in contrast to their American competitors who are still running relay races with their sequential approach to R&D, production and marketing activities. The results of Japanese integrated management techniques in managing R&D, production and marketing have been shown to have shorter lead times (Mansfield, 1988; Clark and Fujimoto, 1989; Graves, 1987, 1992). These lead times have been achieved very often whilst improving product quality to a higher level than that of competitors (Grupp and Hofmeyer, 1986; Womack *et al.*, 1990). Closer integration of R&D with production means that process innovation can be intimately related to product innovation and the joint design of product and process may be the main achievement of Japanese techniques of innovation management. Freeman (1987) has suggested that this may be traced to the experience of reverse engineering in Japan in the 1950s and 1960s when Japanese firms were making extensive use of imported technology, but always with a view to improving it, rather than simply copying. It is notable that in the 1980s US firms still rated reverse engineering as the second most important form of technological learning after in-house R&D (Levin *et al.*, 1987).

It should not be forgotten that talented R&D managers such as Morton (1971) of Bell Labs had already in the 1960s formulated a strategy designed to link R&D closely with production. Nevertheless, Japanese managerial techniques, both for external networking and for internal coupling, as well as for incentives, industrial relations, training and other aspects of their affairs (Dore, 1973, 1985, 1987), have led some economists (notably Aoki, 1986; 1988; 1990; 1991) to regard the Japanese firm as a specific form of industrial organisation (the 'J-Firm'), differing in many important characteristics from American or European firms. Goto (1982) had suggested already in 1982 that the Japanese firms could be regarded as a special type of networking corporation transcending both markets and hierarchies and he noted that in a footnote Williamson (1975) himself had described the Japanese corporate structure as 'culturally specific'. Goto, however, maintained that the advantageous features of Japanese organisation could be diffused internationally. Management consultants, as well as some economists are certainly making considerable efforts to achieve this (Ohmae, 1990).

The importance of learning from production and from marketing, as well as from R&D (Mowery, 1980; Cohen and Levinthal, 1989) help to explain why the 'hollow corporation' has seldom materialised and why subcontracting of R&D and design has not been more extensive. Mowery (1980, 1983) showed historically how and why US firms relied increasingly on their own in-house R&D, rather than on contract Research Institutes, which were already quite strong in the United States at the close of the nineteenth century (Hughes, 1989). Transaction costs are an insufficient explanation: even in industries where contracting for R&D and licensing of technical know-how are common practice, these are hardly ever alternatives to in-house technical activities (including R&D) but are *complementary* to them (see also Reich, 1985). Indeed, one of the most important findings of neo-Schumpeterian research is the demonstration that technical knowledge can seldom be obtained 'off the shelf' and that it almost always requires processing and modification to be used effectively (Bell and Pavitt, 1992; Bell, 1991). Without this assimilation and improvement rather inefficient results are likely to follow, especially in developing countries (Cooper, 1973, 1974; Cooper and Sercovitch, 1971; Bell, 1984; Bell et al., 1976).

In any case, it is not simply a question of in-house R&D, whether in Third World countries or OECD countries. The use of R&D measures as a proxy for a wider range of technical and learning activities is unsatisfactory (Winter, 1987; Bell, 1991; Freeman, 1992). It became common practice in neo-Schumpeterian literature and applied research simply because R&D statistics became generally available for firms, industries and countries during the 1950s and 1960s and were standardised internationally through the OECD in the so-called 'Frascati' Manual (1963, 1970, 1976, 1981, 1993). The authors of this Manual always recognised the limitations of R&D measurements and in the very first edition (1963) pointed to the need to measure a much wider range of scientific and technical services, including design and engineering, project surveys, scientific and technical information services, technical consultancy, training, etc. (see also UNESCO, 1969). In some industries and in industrial countries R&D measures are a reasonably good surrogate for this wider range of activities but in others they are not. A major task for neo-Schumpeterian researchers is to improve their measures of all these technological services (Freeman and Oldham, 1992).

A particularly important aspect of knowledge accumulation is skill formation within firms as the combined result of formal training processes and learning by doing, using and interacting (Marsden, 1993; Kelley and Brooks, 1991). Training statistics have lagged woefully behind but even with the limited data available valuable international comparative studies have been made of the skills of the labour force in Britain, Germany and other countries (e.g., Prais, 1981, 1987; Prais and Wagner, 1983, 1988) and of skill formation in specific industries (e.g., Senker et al., 1985; Brady, 1986).

In his classic study of technical change, Salter (1960) pointed to the enormous variations in productivity between different firms in the same industry and attributed this primarily to different vintages of capital equipment. The neo-Schumpeterians (Dosi, 1984, 1988) have taken on board the central importance of this finding and of firm-specific knowledge accumulation (Penrose, 1959). In discarding the unreal hypothesis of 'representative agents' and recognising heterogeneity of agents, they have also demonstrated the crucial importance of interactive learning in all functions of the firm and of 'core competences' (Teece, 1982, 1987, 1988) in using new vintages of capital equipment. Marshall (1890) already pointed out that knowledge and organisation were the most important forms of capital and the 'chief engine' of production but neither he nor the neo-Schumpeterians have been able to measure this crucially important indicator.

Winter (1987) is one of the few economists who have addressed this problem. In his paper on 'Knowledge and competence as strategic assets', he pointed to the weakness of surrogate measures of knowledge accumulation, using R&D expenditure statistics and argues that it is necessary 'to confront the difficulties that arise from the complexity and diversity of the phenomena denoted by such terms as knowledge, competence, skill and so forth. When we use such terms we hardly ever know precisely what we are talking about . . .' (p. 170). He goes on to discuss the role of tacit knowledge, procedural skills, articulable knowledge and tacit skills and suggests a taxonomy, which could be useful in strategic management. However, neither he nor any other neo-Schumpeterian economist has gone much beyond this. In this area and the related area of measuring scientific and technical services, there is a big challenge to future research, as recognised by several authors (e.g., Coombs, Saviotti and Walsh, 1992; Senker, 1993).

5 Industry-specific incremental and radical innovations

One of the main difficulties in meeting Winter's challenge to neo-Schumpeterian economists lies in the sheer variety and complexity of innovations. They vary enormously by industry (Nelson and Winter, 1977; Nelson, 1984, 1985, 1991; Pavitt, 1984), by degree of novelty and cost (OECD, 1992a), by technology (Dosi, 1982, 1988) and by type (product, process, organisational, system). This section will deal with the first two categories and the other two will be dealt with in sections 6 and 7.

The difficulties of defining and classifying novelty are immense as every patent system is constantly demonstrating. Nevertheless, the very word itself shows the necessity of making some distinctions. An attempt has been made to rank innovations on a five-point scale from 'systemic' to 'major', 'minor', 'incremental' and 'unrecorded' (Freeman, 1971); Abernathy and Clark (1985) used four categories but the vast majority of authors make a simple two-fold distinction between 'radical' (or 'major') innovations and 'incremental' (or 'minor') (e.g., Stobaugh, 1988). Some make no distinction at all. The difficulties of definition are considerable even for this simple dichotomy, but nevertheless it is an important one because the two types of innovation embody a very different mix of knowledge inputs and have very different consequences for the economy and the firms which make them (see especially Utterback, 1993).

Hollander (1965) in his study of technical change in Du Pont's rayon plants and Townsend (1976) in his study of technical change in the British coal industry found that the great majority of innovations did *not* come from formal R&D (even in organisations like Du Pont and the NCB, which had strong in-house R&D facilities). Most of the hundreds of small improvements to the equipment and the organisation of work came from production engineers, systems engineers, technicians, managers, maintenance personnel and, of course, from production workers. Just as Adam Smith (1776) had described, other improvements came from the makers of plant and machinery. In this respect both Du Pont and the NCB could be described as 'users' and as we have already seen many other studies have amply confirmed the role of users in making incremental innovations (e.g., Rothwell, 1977; Cassiolato, 1992; Lundvall, 1985, 1988; Lundvall, ed., 1992; von Hippel, 1978, 1988; Slaughter, 1993). However, when it came to a change in the basic process in the rayon industry or to the introduction of electronic 'black boxes' in coal-mining (i.e., a *discontinuity* in products and processes), then central R&D departments, whether in user or producer firms, became very important, because the new knowledge required went beyond the experience of the people involved in production (see also Utterback, 1979, 1993; Afuah and Utterback, 1991).

This points the way to the definitions proposed by Mensch (1975), Utterback (1979) and Freeman and Perez (1988). Mensch defines a radical innovation as one that needs a new factory and/or market for its exploitation and this is similar to Utterback's definition of major or radical innovations. Freeman and Perez added to this the suggestion that, logically, radical innovations would need a new column and a new row in a complete input–output table. Incremental innovations, on the other hand, would need only new coefficients in the table of existing products and services since they refer only to improvements in the existing range of output. It must be noted that the *discontinuity* with radical innovations is in *production and marketing* systems and not necessarily in the *firm* (Pavitt, 1986a), although some authors also emphasise the contribution of new firms to radical innovation, as in the US semi-conductor industry (e.g., Tilton, 1971; Braun and MacDonald, 1978; Saxenian, 1991).

If one moves away from improvements in production to entirely new products and processes, whether in the chemical industry or in the coal industry, then more formal inputs from the science–technology system are usually needed, which, in contemporary industrialised economies, usually means from the R&D system (Dosi, 1988). Such radical innovations are much more completely recorded than incremental innovations. They are often the subject of papers in technical journals and of manuals and textbooks as well as blue-prints, specifications and patents, and hardware prototypes and pilot plants (Ames, 1961; Machlup, 1962).

Neo-Schumpeterian research has certainly given us a much more complete picture of what is involved in radical innovation in many different industries. However, it has not yet succeeded in developing regular national and international *statistics* of innovations, despite pioneering attempts, which are well described in the *STI Review* (OECD, 1992b) with papers on innovation surveys in six European countries.

Studies in particular *sectors*, such as synthetic materials (Hufbauer, 1966), pesticides (Achilladelis, Schwarzkopf and Lines, 1987) or particular countries (Pavitt, Robson and Townsend, 1987; OECD, 1992b) have shown the value of more complete records and data banks and of long-term historical perspective. However, the absence of sufficient good historical investigations means that the results of some old controversies are still inconclusive and the answers to some important questions are not yet available. One example is the controversy about the supposed clustering of radical innovations at certain phases of long cycles in economic development (Mensch, 1975; Kleinknecht, 1987, 1990; Clark, Freeman and Soete, 1981; Freeman, Clark and Soete, 1982; Freeman and Perez, 1988). Mensch and Kleinknecht in their respective works made use of various independent sources on the history of inventions and innovations but neither of them could be described as comprehensive. Another very difficult question is the relative contribution of various national systems to world-wide innovation in different historical periods. Dosi, Pavitt and Soete (1990) made use of various incomplete records to provide a partial answer but again they would be the first to agree that these data are far from satisfactory.

Radical innovations played such an obvious and spectacular role during and after the Second World War (especially of course, nuclear weapons and radar) that they tended to overshadow incremental innovations, both in policy making and in descriptive analysis for a long time. Neo-Schumpeterian research has succeeded in restoring a much more balanced picture of the overall process of technical change both by the indirect evidence of productivity studies (OECD, 1991b, 1992a) and by direct studies, such as those by Hollander and Townsend (see also Surrey, 1992). This has led to an important change of emphasis in policy making with the recognition that the vast majority of firms do not make radical innovations, but all can and should make incremental innovations and adopt new products and processes first made by others (see section 6). This shift of emphasis towards diffusion can be observed in most OECD countries in the 1980s and 1990s (OECD, 1991a; 1992a, Ergas, 1984; Arundel and Soete, eds., 1993).

Even though many incremental innovations go unrecorded and unarticulated in the general process of tacit knowledge accumulation, learning by doing, using and inter-acting (see especially Gilfillan, 1935; Utterback, 1979; Lundvall, ed., 1992), it has proved possible to document and measure many aspects of incremental innovations. The *recorded* part of incremental innovation can be traced through technical journals, business histories and above all through patent statistics, which represent a unique long time series of inventive efforts on a world-wide basis. Schmookler (1966) especially pioneered the systematic use of patent statistics in the study of technical change and even though some of his conclusions are highly controversial (see section 6) he inspired many others to make use of this gold-mine for empirical research. More recently computerised patent systems have greatly facilitated part of this work.

It is true of course that patents refer to the outcome of *inventive* efforts and are therefore not a direct measure of *innovations*. Nevertheless, they are clearly a very important intermediate output of innovative activities and even though many patents lapse, there is a considerable overlap especially with incremental innovations. It is also true that there are hazards in the use of raw patent statistics (Scherer, 1983; Basberg, 1987; Pavitt, 1982, 1985; Griliches, 1990a; Grupp, 1991; Grupp, ed., 1992), such as varying national legal systems, varying propensities to patent, industry and technology-specific features, classification systems and so forth. Nevertheless, neo-Schumpeterians have on the whole shown admirable caution and much ingenuity in coping with these problems and are still producing worthwhile results, using patent statistics as a surrogate measure of innovative output (e.g., Schmookler, 1966; Reekie, 1973; Soete,1981; Archibugi, 1988a, 1988b; Archibugi, Cesaratto and Sirilli, 1987; Archibugi and Pianta, 1992; Soete and Wyatt, 1983; Fagerberg 1988; Scherer, 1982a, 1982b, 1986, 1992; Cantwell, 1989, 1991b; Narin *et al.* 1985, 1987, 1992; Sirilli, 1987; Freeman *et al.* 1963, 1987; Freeman, ed., 1987; Grupp, 1992; Patel and Soete, 1988; Patel and Pavitt, 1991a, 1992a, 1992b, 1995; Griliches, 1990a; Griliches, ed., 1984; von Tunzelmann, 1989).

Far less use has been made of other sources, which though less accessible and less comprehensive than patent records are nevertheless potentially valuable also. Townsend (1976) is one of the few authors who has made systematic use of data on inventor awards, although some firms have used such schemes for long periods. A difficulty with some of these measures is that they may be oriented more towards industrial relations and public relations than towards innovation. However, now that Japanese firms have been shown to promote incremental innovation rather successfully by a variety of incentive schemes including inventor awards, quality circles, etc. (e.g., Aoki, 1988; Baba, 1985; Imai and Itami, 1984; Dore, 1973, 1985) greater attention is being paid to this topic. Bessant and his colleagues have initiated a research programme specifically to study 'continuous innovation' (Bessant *et al.*, 1993; CI News, 1993) (See also Petroski, 1989 and Paulinyi, 1982, 1989 for continuous innovation in the nineteenth century).

The distinction between radical and incremental innovations is also relevant in terms of the specific features of industrial sectors. Only a few firms make radical innovations and they are clustered in certain industries. There are many service industries and some manufacturing industries which make scarcely any radical innovations, and some which make none at all. As pointed out in section 1, Nelson and Winter (1977) stressed particularly the importance of these interindustry variations and there has been a considerable upsurge of industry-specific studies since that time. It is true that most of this research has concentrated on the more 'glamorous' and R&D-intensive industries and mostly though not exclusively on radical innovations, especially in electronics (e.g., Nelson, 1962; Braun and Macdonald, 1978; Sciberras, 1977; Freeman *et al.*, 1965; Dosi, 1984; Malerba, 1985; Malerba *et al.*, 1991; Baba, 1985; Flamm, 1987, 1988; Molina, 1989, 1990; Lovio, 1993; Katz and Phillips, 1982; Sternberg, 1992; Lundgren, 1992; Grupp and Soete, 1993; Antonelli, 1992, 1993), in chemicals (e.g., Enos, 1962; Hounshell, 1992a, 1992b; Hounshell and Smith, 1988; Morris, 1982; Stobaugh, 1988; Achilladelis, Schwarzkopf and Lines, 1987, 1990; Quintella, 1993; Walsh, 1984) and in nuclear power (e.g., Gowing, 1964; Surrey, 1973; Keck, 1982; Samuels, 1987; Walker and Lönnroth, 1983a, 1983b; Cowan, 1990; MacKerron, 1991; Surrey and Thomas, 1980; Krupp, 1992; Thomas, 1988).

Already in the 1950s and 1960s, pioneering studies by Rosenberg (1963) and other historians had investigated machine tools and these less research-intensive industries such as automobiles and machinery have recently begun to receive much more attention (e.g., Womack, Jones and Roos, 1990; Dankbaar, 1993; Ayres, 1991a, 1991b; Graves, 1992; Bessant, 1991; Tidd, 1991; Jaikumar, 1988; Jacobsson, 1986). Even industries of very low research intensity, such as textiles (e.g., Antonelli, Petit and Tahar, 1992), construction (Gann, 1992, 1993) and clothing (Hoffmann and Rush, 1988; Whittaker, Rush and Haywood, 1989) are becoming the object of more systematic research.

A serious weakness of this research has been the lack of attention to service industries but this is also a criticism which could be made of industrial economics more generally. Nevertheless, a few pioneering studies have been made, such as Auliana Poon's (1993) highly original research on technical change in tourism, and some work on financial services (e.g., Heertje, ed., 1988; Baba and Takai, 1990; Petit, 1991; Wit, 1990; Cassiolato, 1992; Christensen, 1992). These studies together with more general analysis of service industries (Barras, 1986, 1990; Quinn, 1986; Posthuma, 1986) show that most of them shared some of the characteristics of manufacturing industries of very low research intensity (Scherer, 1982b) but are now changing.

It was already evident from the early R&D statistics in the 1950s that industries fall into quite distinct categories (Nelson, ed., 1962; Mansfield, 1968; Mansfield *et al.*, 1971; Freeman, 1962) of high, medium and low R&D intensity and that the same industries belonged to each category, in whichever country they were located, suggesting that technological trajectories and opportunities and technological competition had a dominant influence on this aspect of firm behaviour (see also Kay, 1979, 1982). These early studies also showed that the fastest growing industries were mainly those with the highest R&D intensity.

A more systematic taxonomy of industrial sectors was developed by Pavitt (1984) which has been widely used in neo-Schumpeterian literature. He classified industries into three categories: (a) supplier dominated, (b) production intensive, and (c) science based. Pavitt argued that each category had a somewhat different pattern of external relationships to sources of knowledge, of in-house scientific and technical activities, of diversification behaviour, of industrial structure and of skill formation. This taxonomy proved a fruitful framework for analysis, further justifying Nelson and Winter's insistence on the importance of intersectoral variations both for theory and for policy making. Scherer (1982a) had already demonstrated the importance of interindustry technology flows by the use of patent statistics. Utterback (1993) makes an interesting distinction between assembly and non-assembly industries with respect to radical and incremental innovations.

However, the example of the service industries also shows the dangers of classification schemes and taxonomies for neo-Schumpeterians. It was customary and reasonable for a long time to regard the service industries as 'supplier-dominated' in the sense that other industries (office machinery, telecommunications, etc.) were responsible for introducing and diffusing most of the technical change which was taking place. With the computer revolution this may be changing. In-house software development (not always classified as R&D – another problem for neo-Schumpeterians!) is now characteristic of many firms in financial services, who also have a heavier investment in ICT equipment than most firms in manufacturing. At the same time specialist software companies are proliferating and have a very dynamic role in technical change (Quintas, ed., 1993; Cusumano, 1991; Brady and Quintas, 1991; Freeman, 1993). Neo-Schumpeterians have to beware always of 'freezing' a classification or a theory, which may be out-dated by the incessant changes in technology (Barras, 1986, 1990; Poon, 1993; Bressand and Kalypso, eds., 1989) and in industrial structure. Miller *et al.*, (1993) have suggested that complex systems such as aircraft flight simulators are made and sold by firms which do not fit into any of Pavitt's categories.

The same danger exists in relation to the old Schumpeterian controversy about the role of oligopoly in innovation. Schumpeter's emphasis on the dominant role of large oligopolistic firms was challenged by many authors (see Kamien and Schwartz, 1982; Hamberg, 1964, 1966; Scherer, 1980). Jewkes *et al.* (1958) in particular attempted to show from the empirical evidence of case studies that major *inventions* were made as frequently by small firms and individual inventors in the twentieth century as in the nineteenth, and that the contribution of large firms was exaggerated (see also MacQueen and Wallmark, 1983). As they themselves emphasised, their findings related to *inventions*, not to innovations. Their own case studies show that about two thirds of their inventions were actually *innovated* by large firms (Freeman, 1992) and more frequently in the twentieth than in the nineteenth century.

However, more recent empirical work has still confirmed their main finding that small firms do indeed continue to make a significant contribution both to invention and innovation and that it may have recently increased (Pavitt, Robson and Townsend, 1987; Acs and Audretsch, 1988). Large firms still account for the great majority of innovations in most industries and the contribution of small firms tends to be concentrated in a few industrial sectors, but their innovations may be used in many others (Pavitt, 1984). The instrument industry and the software industry are obvious examples.

Here too, however, history matters. As Kaplinsky (1983) and others (e.g., Oakey, 1984) have shown in relation to CAD and many other innovations associated with new generic technologies (especially bio-technology) in the early stages, small new firms make an outstanding contribution. However, as the technologies mature, R&D costs often rise, acquisitions take place and Schumpeterian competition may lead to renewed concentration (Utterback and Suarez, 1993). It remains to be seen whether 'new history' will lead to a more stable symbiosis of large and small firms in networks of technological collaboration, as foreshadowed by some authors (section 3) or to renewed concentrations or to de-concentration. It should be noted that the contribution of small firms is exaggerated in some reports through failure to distinguish 'small' subsidiaries of MNCs and establishments from firms, and innovations which are new to the firm from those which are new to the world. Longitudinal and historical studies in a variety of industries (e.g., Lundgren, 1992; Lovio, 1993; Utterback and Suarez, 1993; Utterback, 1993) distinguishing the role of various size and ownership categories and of networks will be a vitally important task for future neo-Schumpeterian research.

6 Demand-pull, technology push and diffusion of innovations

The discussion in sections 3, 4 and 5 has brought out the importance of flows of information and knowledge between firms as well as within firms.

Moreover, the results of the empirical research which has been described (especially in section 3), point to the importance both of flows to and from sources of scientific and technical knowledge and of flows to and from users of products and processes. For a long time controversy has raged both among economists and among historians of science and technology about the relative significance of 'demand-pull' versus 'science and technology push' in generating and sustaining these flows (e.g., Hessen, 1931; Schmookler, 1966; Bernal, 1939, 1970; Scherer, 1982b; Verspagen and Kleinknecht, 1990).

The distinction between radical and incremental innovation is highly relevant here as is the pattern of diffusion of an innovation. In the early stages of a truly radical innovation, scientific and technological inputs are likely to be prominent, even if they do not provide the original impulse. Katz and Phillips (1982) have shown that in the early days of the electronic computer (arguably the most important twentieth century innovation) science and technology push predominated and even industrialists as knowledgeable as T.J. Watson (Senior) maintained there was and would be no market demand. Other studies (e.g., Molina, 1989, 1990) have shown that the science 'constituency' was also prominent in originating and shaping later radical innovations in the computer industry. Sociologists (e.g., Mackenzie, 1990a, 1990b; Bijker and Law, eds., 1992) have made a major contribution here (see Coombs et al., 1992).

Nevertheless, in the 1960s and 1970s demand-led theories of innovation made a considerable impact on policy makers. The empirical survey of over 500 innovations made by Myers and Marquis (1969) appeared to justify the demand-pull approach whilst on a more theoretical level the work of Schmookler (1966) provided a more sophisticated historical justification. He did not entirely deny the independent role of basic scientific research but sought to demonstrate through the painstaking use of very detailed US patent statistics that usually the peaks and troughs of *inventive* activity lagged behind the peaks and troughs of *investment* activity. From this he drew the conclusion that the main stimulus to invention and innovation came from the changing pattern of demand as measured by investment in new capital goods in various industries.

Scherer (1982b) tested Schmookler's hypothesis for a more comprehensive set of US manufacturing industries and found a much weaker relationship. Verspagen and Kleinknecht (1990) also found Schmookler's own data showed a weaker relationship than he had claimed. Scherer made an original analysis of cross-sector flows of patent origin and use which demonstrated that the link for inventions sold across industry lines was at least as strong as for those that represented internal processes to their originators.

Pure demand-pull theories were already strongly criticised in the 1970s (e.g., Rosenberg, 1976). However the *coup de grâce* was given by Mowery and Rosenberg (1979) in their devastating review of 'Market demand and innovation'. They showed that empirical studies of innovation which were often cited in support of 'demand-pull' did not in fact justify these conclusions and indeed that the authors themselves repudiated this interpretation (e.g., Langrish *et al.*, 1972; Freeman, 1974). Mowery and Rosenberg further pointed to the confusion in the literature between 'needs' and 'demand' and between 'potential demand' and 'effective demand'. Since human 'needs' are extremely varied and often unsatisfied for long periods, they cannot alone explain the emergence of particular innovations at a particular time. Innovation should not be viewed as a linear process, whether led by demand or by technology, but as a complex interaction linking potential users with new developments in science and technology.

The majority of innovations characterised as 'demand led' in the Myers and Marquis survey were actually relatively minor innovations along established trajectories and the same was true of the vast majority of patents analysed by Schmookler. The Mowery and Rosenberg critique of demand-pull has been further reinforced by research using patent statistics and Schmookler's own method (Walsh, 1984). In her study of the chemical industry, Walsh made use of statistics of scientific papers as well as patents and related these to measures of output, investment, innovation and sales after the manner of Schmookler. As in Schmookler's work the pattern of leads and lags was by no means so clear-cut as to put the issue beyond all doubt. But as with his work there was evidence of synchronicity in the pattern of economic and technical developments i.e., the major upsurge of production and investment in each sector (petro-chemicals, dyestuffs, drugs and synthetic materials) was accompanied by a remarkable increase both in the numbers of patents and in the output of related scientific publications. The most interesting result, however, was the evidence suggesting that a 'counter-Schmookler' pattern was characteristic of the early stages of innovation in synthetic materials, drugs and dyestuffs, changing to something more closely resembling a 'Schmookler' type pattern once the industry 'took off'. Qualitative analysis in all four cases confirmed the importance of early scientific and technological breakthroughs permitting and triggering an upsurge of inventive activity and technical innovations. The work of Fleck (1983, 1988) on Robotics shows a similar pattern of early science-technology push, followed by numerous system improvements in specific applications driven by users interacting with suppliers.

These results of empirical research point to the resolution of the persistent controversy between adherents of 'demand-pull' or 'market-led' theories of innovation and advocates of 'technology push' or 'science-driven' theories. One of the achievements of innovation research has been to bury linear models of innovation, whether supply or demand driven and to replace them with more sophisticated models (Arundel and Soete, eds., 1993; Rothwell, 1991; OECD, 1992) which embody the numerous interactions and feedback loops during both innovation and diffusion.

Diffusion of innovations is another Schumpeterian concept which has come in for some heavy criticism. Most empirical studies demonstrate that new products and processes are usually changed considerably during the diffusion process. The early models of the 1950s and 1960s which tended to assume an unchanged product diffusing through an unchanged environment have been largely displaced by more complex models (see Metcalfe, 1981, 1988; Mahajan and Peterson, 1979; Gold, 1981; Davies, 1979; Zuscovich, 1984; Mansfield, 1989; Mansfield *et al.*, 1977; Stoneman, 1976, 1983, 1987; Granstrand, 1986; Nakicenovic and Grübler, eds., 1991; Midgley *et al.*, 1992; Callon, 1993 and see also references to David and Arthur in section 2).

It is impossible to deny the evidence of further innovations during diffusion and in many cases productivity gains are mainly due to this learning process and the competitive pressures engendered by band-wagon effects. Rosenberg (1976, 1982) in particular, has always insisted on this point which could hardly be put more forcefully than by Kline and Rosenberg (1985):

It is a serious mistake to treat an innovation as if it were a well-defined homogenous thing that could be identified as entering the economy at a precise date. . . . The fact is that most important innovations go through rather drastic changes over their life-times – changes that may, and often do, totally transform their economic significance. The subsequent improvements in an invention after its first introduction may be vastly more important, economically than the initial availability of the invention in its original form ... consider the performance characteristics of the telephone around 1880, the automobile, vintage 1900, or the airplane when the Wright Brothers achieved their first heavier-than-air flight in 1903 – in that form, at best a frail and economically worthless novelty.

One could add to these examples that of the electronic computer, where contemporary micro-processors and super-computers can barely be classed as the same product as Zuse's first electronic computer at Charlottenburg Technische Hochschule in 1940 (Freeman *et al.*, 1965; Stoneman, 1976; Flamm, 1987). To be fair to Schumpeter he did recognise this point. He said himself that:

the motor car would never have acquired its present importance and become so potent a reformer of life if it had remained what it was thirty years ago and if it had failed to shape the environmental conditions – roads among them – for its own further development. (Schumpeter, 1939, p. 167)

These factors led Fleck (1988) to coin the expressions 'Innofusion' and 'Diffusation' in his analysis of industrial Robotics. However, although it is certainly possible to identify products which have been drastically modified and improved during their diffusion (such as the automobile or the computer), there are also some which change little or not at all. Examples can be found particularly in the drug industry and the food industry. Even when the product or process is improved during diffusion, this improvement is often very gradual and can be legitimately regarded as *incremental* in nature, rather than *radical*. The pioneering studies of Mansfield (1961) and other researchers of the 1960s contributed a great deal to our understanding of diffusion despite their limitations.

Thus, despite the criticisms of Schumpeter's theory of diffusion, most social scientists still find his conceptual framework useful. Economists and sociologists are actually studying diffusion more than ever, although they are increasingly aware of the need to take into account that they are often examining a changing product in a changing environment. The Venice Conference on diffusion (Arcangeli, David and Dosi, eds., 1986) was the biggest event of its kind for neo-Schumpeterian economists between the NBER Conference in 1961 and the MERIT 5th Anniversary Conference in 1992. However, despite the recognition that diffusion has been rather neglected – described by Paul David (1986a) as the 'Cinderella' of technology policy – diffusion policies in OECD countries remain inadequate, and still lack satisfactory criteria and guideposts (OECD, 1991a, 1991b).

One of the major problems in diffusion research is to take into account the *supply* side as well as the demand side. As Gibbons and Metcalfe (1986) particularly show the interaction between supply and demand results in the evolution both of new and improved products and of new design configurations. Although empirical research has amply demonstrated the role of suppliers in improving the product, diversifying new models, enlarging the market, promoting applications research, training potential users and coping with institutional barriers, much diffusion research (despite the good advice of Metcalfe, 1988 and other researchers) continues to neglect the supply side and treat diffusion as a demand phenomenon. The lessons of the demand-pull/technology-push debate have still not been fully assimilated. The model of 'network structure' in industrial diffusion processes (Midgley *et al.*, 1992) is an interesting example of the new type of simulation model which does endeavour to take into account the interactions of suppliers and adopters, as well as some third parties. A particularly interesting finding of this research is that third parties may be as important to the diffusion process in some areas as direct links from adopters to potential adopters or suppliers to adopters.

Another even more difficult problem is the systems aspect of diffusion.

Most innovations are not discrete events or isolated products but form part of a technology system (Gille, 1978; Hughes, 1982, 1992; Carlsson and Stankiewicz, 1991). Obvious examples are contemporary electronic products which require appropriate software, user-friendly peripherals and so forth. Diffusion research on computer-controlled machine tools and robotics (Ayres, 1991b; Camagni, ed., 1991; Camagni *et al.*, 1984; Bessant and Haywood, 1991; Bessant, 1991; Jacobsson, 1986; Carlsson and Jacobsson, 1993; Edqvist, 1989; Fleck, 1983, 1988, 1993; Arcangeli *et al.*, 1991) has demonstrated the enormous importance of these systemic features and of local skills and institutions but it cannot yet be said that international comparisons of diffusion of FMS (flexible manufacturing systems) and still less CIM (computer integrated manufacturing) yet give us a full picture of what is happening on a comparative international basis. The problems of compatibility standards also play a complex role in the diffusion of systems (Greenstein, 1990; David and Steinmueller, 1990; David and Greenstein, 1990).

Diffusion research in this and other fields has nevertheless shown conclusively that rates of diffusion vary a great deal, both by product, by system and by country (Nabseth and Ray, 1974; Ray, 1984; Romeo, 1975; Nakicemovic and Grübler, eds., 1991; Jacobsson, 1986; Rosenberg, 1976; Stoneman, 1976, 1983; Dosi, 1991; Mansfield, 1989; Ayres, 1991a). It has also shown that the productivity gains associated with diffusion vary enormously partly because of the strong systemic features of most innovation (see Hughes, 1982, David, 1991, for electric power; Antonelli, 1993, 1986; Mansell and Morgan, 1991; Miles, 1989; Jagger and Miles, 1991; Thomas and Miles, 1989, for telecommunications and telematic services; Bessant and Haywood, 1991; Jacobsson, 1986; Fleck, 1983, 1988, 1993; Kodama, 1986, 1991, 1992 for mechatronics). Moreover, comparative international research (e.g., Nabseth and Ray, 1974; Ray, 1984) has also shown that the country of first innovation is not necessarily the same as the country of most rapid diffusion or highest productivity gains.

7 Institutional change, trajectories and paradigms

In attempting to explain interfirm, interindustry and intercountry differences in rates of diffusion and associated productivity gains, neo-Schumpeterians have shown that such differences cannot be attributed simply to capital-embodied technical change but are heavily dependent in the first place on skill intensity, learning and training (as discussed in section 4) and on managerial and organisational innovations in such areas as labour relations, incentives, hierarchical managerial structures, communication systems within and between firms, stock control systems and so forth (e.g., Matthews, 1989; Dertouzos *et al.*, 1989; OECD, 1991a, 1991b, 1992a; Whiston, 1989; Perez, 1989; Womack, Jones and Roos, 1990; Sorge, 1993; Piore, 1993; Sorge *et al.*, 1990; Cressey and Williams, 1990; Watanabe, 1993; Bessant and Haywood, 1991; Gjerding *et al.*, 1992; Dankbaar, 1993). This is especially true when it is a question of new technological systems (Bailey and Chakrabarti, 1985; David, 1991). Historians, such as Lazonick (1990, 1992a, 1992b; Chandler, 1977, 1990; Landes, 1970), had of course demonstrated the enormous importance of managerial innovations and their connections with more narrow technical innovations in earlier waves of technical change. There are welcome indications of increasing coopera-tion and convergence between economists, business historians, organisa-tion theorists and sociologists in studying these phenomena (see, e.g., Coombs, Saviotti and Walsh, 1987, 1992, chapter 1; Nakicenovic and Grübler, 1991; Dosi *et al.*, eds., 1992, and the new journal *Industrial and Corporate Change*).

By definition any radical innovation involves some change in the organ-isation of production and markets. Organisational and institutional inno-vations are thus inextricably associated with technical innovations. They are also often related to changes in infrastructure as Kondratieff and Schumpeter suggested (see the pioneering study by Grübler, 1990). Moreover, as Schumpeter also pointed out, sometimes organisational and managerial innovations can lead the way. Obvious major examples are the assembly line, containerisation, self-service, supermarkets and hypermar-kets. Organisational innovations may often induce technical innovations too (Klein, 1977) but do not necessarily do so. On the other hand, techni-cal developments in ICT have greatly facilitated new telematic services, such as telematic shopping, telematic banking and data networks (Miles, 1989, 1990; Thomas and Miles, 1989; Mansell, 1988, 1989; Jagger and Miles, 1991; Miles *et al.*, 1988) which may transform many aspects of social life (see also Gershuny, 1983; Gershuny and Miles, 1983 for social innova-tions in the division of labour). The introduction, implementation and diffusion of such generic technologies as information and communication technology or biotechnology is so obviously and intimately associated with a complex process of institutional and infrastructure change that neo-Schumpeterians (like the management of the firms which they study), have been obliged to pay more and more attention to these issues (Perez, 1985, 1989).

Another reason for the increasing preoccupation of neo-Schumpeterian economists with institutional change lies in the need to explain how rela-tively ordered processes of technical change can emerge from the diversity and uncertainty associated with invention and innovation. As evolutionary economists such as Boyer (1988, 1993), Eliasson (1988), Dosi and Orsenigo

(1988), Johnson (1992), Foray (1993) and Nelson and Winter (1982) have strongly insisted, while macroeconomic growth requires microeconomic diversity and instability, it also requires processes of harmonisation, regulation, standardisation and routinisation to avoid chaotic instability and to reap the benefits of scale economies. In studying these processes neo-Schumpeterians have been responding to another of the challenges posed by Nelson and Winter (1977) in their search for 'useful theories of innovation'. They themselves (1982) assigned an important role to behavioural routines and also suggested that technologies had 'natural trajectories' of their own which enabled designers, engineers, managers and entrepreneurs to visualise likely future paths of development and growth. Very pervasive technologies, such as electric power with multiple potential applications might be described as 'generalised natural trajectories'.

The notion of 'natural' trajectories has on the other hand been criticised by some sociologists (e.g., Mackenzie, 1990b; Sorge, 1993) on the grounds that these trajectories are not 'natural' phenomena but are socially and institutionally determined by the expectations, theories, activities and self-interest of engineers, scientists, managers, entrepreneurs and their 'constituencies' (Bijker and Law, eds., 1992). The 'natural' rate of interest and 'natural' rate of unemployment could be criticised on much the same grounds (Freeman, 1992). Nevertheless, the critics do not deny that trajectories, whatever their basis, do play an important role in the evolution of technology and do impart a definite pattern of development in much the same way as Kuhn's (1962) 'normal' science. A useful and influential analogy has been made between scientific and technological paradigms (Dosi, 1982).

However, even within a paradigm, by definition innovation involves initially an increase in diversity – an extension of the range of products, processes and services. In the early stages of *diffusion*, also, as the research reviewed above has shown, this diversity will usually increase. Burton Klein (1977) has described the wide range of steam cars, electric cars and petrol-driven cars available in the United States in the early days of the internal combustion engine. Fleck (1983, 1988) has described the diversification of design of industrial robots to meet the wide variety of application environments. Utterback (1993) has analysed several industries and shown similar patterns in the early stages. At this stage, design is fluid, there are no standards (or very few) and there is great uncertainty about the future of the new products. This diversity may be further increased by the divergence of *national* markets, production and technology systems (e.g., Foray and Grübler, 1990). Since radical innovations also involve organisational and other institutional innovations there is scope for very considerable variety. Thus, although the assembly line, introduced by Ford to manufacture automobiles, was imitated in many other countries and other industries, Boyer (1988) describes several variations in Europe as well as in Japan, which made 'Fordism' a somewhat different system in each country. Boyer (1993), Foray (1993), Saviotti (1991) and Johnson (1992) are among many authors who maintain that institutional diversity in the diffusion of new technologies is an essential and beneficial feature of evolutionary development, as well as the technical diversity associated with technological competition (see Part IV of Foray and Freeman, eds., 1993).

It is the institutional diversity which has led to the growing interest among neo-Schumpeterians in what have become known as 'national systems of innovation' (Lundvall, 1985; Lundvall, ed., 1992; Freeman, 1987; Nelson, 1985, 1992a; Nelson, ed., 1993; Mjøset, 1992; Niosi, ed., 1991). It is not only firms which diverge in their mode of implementing new technologies and in their management systems but many other institutions (Hollingsworth, 1993). Since firms depend on a variety of external linkages in acquiring the necessary technical, scientific and organisational knowledge, information and skills (section 3), it is obvious that the national education and training system may have a considerable influence on firms' innovative performance, as well as a wide variety of Research Institutes, sources of technical information, consultancy services and government laboratories. However, most neo-Schumpeterians, following Lundvall (ed., 1992) and his colleagues stress that a 'national system of innovation' is much more than a network of institutions supporting R&D, it involves interfirm network relationships and especially user–producer linkages of all kinds (Andersen, 1992a), as well as incentive and appropriability systems, labour relations and a wide range of government institutions and policies (see also Perrin, 1988). Just as heterogeneity of firms and oligopoly has led neo-Schumpeterians to discard the assumptions of representative agents and perfect competition (Sylos Labini, 1962; Dosi, 1984, 1988), so heterogeneity of national systems of innovation and hegemony of great powers has led them to discard notions of international convergence and to stress the phenomena of divergence in growth rates, 'forging ahead', catching up and falling behind (Abramowitz, 1986; Lundvall, ed., 1992; Dosi and Freeman, 1992; Verspagen, 1992).

Dahmen (1950, 1988) already anticipated some features of 'national systems' with his theory of 'development blocks'. Three recent major contributions to the literature on national systems are Lundvall (ed., 1992), Mjøset (1992) and Nelson (ed., 1993). The Nelson book is a comparative study of a dozen or more countries, while the Lundvall book is a more theoretical analysis. The Mjøset book is about the Irish economy but reviews this in the context of comparisons with other small European countries. Together they go far to explain why countries like the United States (late nineteenth century) (Mowery, 1992b) and Japan (late twentieth century) could catch up with the erstwhile leaders and forge ahead in technology without necessarily leading in basic scientific research at the time (but see also Hicks *et al.*, 1992a and 1992b; and Cantwell, 1993). However, the Nelson Study, together with such in-depth studies of Korea and Taiwan as Wade (1990), Amsden (1989), Jang-Sup Shin (1992), Hobday (1992) show that catching up countries do need the sort of capability in basic science and education insisted upon by Pavitt (1993) as part of their national systems (see also Villaschi, 1992, on the case of Brazil). The work of the neo-Schumpeterians has clearly had a considerable influence on development economics as well as on the so-called 'new growth theory', with the acceptance of the key role of intangible investment in economic development (see especially World Bank, 1991). The research also of course throws much light on why countries fall behind (Walker, 1993; Cantwell, 1991b; Mjøset, 1992).

A major problem confronting this neo-Schumpeterian research tradition is to reconcile their emphasis on national diversity (and regional diversity: see Saxenian, 1991; Storper and Harrison, 1991; Morgan and Sayer, 1988; Bianchi and Bellini, 1991; Russo, 1985; Oakey, 1984; Camagni, 1991; Goddard, Thwaites and Gibbs, 1986; Thwaites and Oakey, eds., 1985; Thwaites, 1978; Harris, 1988; Scott, 1991; Alderman and Davies, 1990; Alderman *et al.*, 1993; Amin and Goddard, eds., 1986) in systems of innovation with the role of multinational corporations (MNCs) in the diffusion and global standardisation of both management methods and technical developments (Cantwell, 1989, 1991a, 1993; Casson, ed., 1991; Dunning, 1988; Chesnais, 1988a, 1992). Following the tendency of multinationals to locate some production outside their home base there has also been a tendency to locate some R&D activities abroad. The extent and consequences of this tendency have been strongly debated among neo-Schumpeterians (Patel and Pavitt, 1991a; Ohmae, 1990; Hu, 1992; Pearce, 1990; Pearce and Singh, 1992; Porter, 1990; Freeman and Hagedoorn, 1992; Howells, 1990; Pausenberger, 1991; Wortmann, 1990; Reich, 1991; Niosi, ed., 1991; Cantwell, 1993). Some have stressed the role of cultural, educational, political and geographical factors (especially in diffusing tacit knowledge), in reinforcing national and regional networks of collaboration (Lundvall, ed., 1992; Porter, 1990). Others have pointed to the role of world-wide liberalisation of trade and investment and of telecommunications and airlines in reducing the significance of such national influences and strengthening globalisation (Ohmae, 1990). Whilst 'globalisation' need not necessarily mean world-wide standardisation and loss of diversity, there are clearly powerful incentives within MNCs to reap scale economies in R&D and design as well as production and marketing.

Some degree of national and international 'lock-in' by standardisation is unavoidable for a whole variety of reasons such as the need for 'connectivity' in systemic innovations, consumer acceptance and learning, as well as for scale economies. All of these have been especially evident in the world of computers and their multiple applications in industry and services, as well as in telecommunication systems and the new network services (Rosario and Schmidt, 1991; Hawkins, 1992). The dilemma for standards of 'freezing' too early and thereby locking out potentially superior innovations at least for a while, or of failing to set standards early enough is one that is permanently associated with innovation and is resolved not by optimal rationality but by oligopolistic competition, political bargaining and conflict at both national and international levels (Rosario and Schmidt, 1991; Mansell, 1990b; Mansell and Morgan, 1991; David, 1986a; David et al., 1990a, 1990b; Greenstein, 1990).

Nor is this dilemma confined to diffusion and standardisation, as Gjerbing (1992) and Holbek (1988) have shown; it is inherent in innovation itself. There is always the possibility of inproving a design and/or a technology so that the firm is constantly involved in the dilemma of when to 'freeze' the design. As Rothwell and Gardiner (1988) have shown the best designs are 'robust designs' which have the potential for 'stretch' and adaptation over a considerable period, but these are robust not optimal (see also Utterback and Suarez, 1993, for the concept of 'dominant design').

Finally, there has always been a dilemma between the choice of incentives for *innovation* and incentives for *diffusion*. The patent system has had as one of its primary objectives to provide a stimulus to invention by granting a temporary monopoly for exploitation but this monopoly may of course retard diffusion. The dilemma of designing an 'optimal' appropriability regime has never been satisfactorily resolved as is evident both from the literature (e.g., Teece, 1987; Taylor and Silberston, 1973; Levin, 1986, 1988; Levin et al., 1985, 1987; Cohendet et al., 1993; Antonelli and Foray, 1991; Foray, 1992; Winter, 1989, 1993; Mansfield et al., 1981) and from the history of patent legislation in many countries. Neo-Schumpeterian research has made a significant contribution in this field, by confirming the relatively limited role that patents play in protecting firms' new technology in most industries (Levin et al., 1987) and by demonstrating the important contribution of other methods. The Yale Survey has also been valuable in demonstrating the industry-specific nature of these methods; the extension of this Survey to Europe in 1993 is therefore very welcome.

The constant re-emergence of novelty and diversity means that unique features of historical development must always be important for neo-Schumpeterians (see, for example, Pierre Dockès on 'histoire raisonée et économie historique', 1991). History always matters and much research has been devoted to 'mapping' the nature and direction of technical change. Sahal (1977, 1981, 1985), Dosi (1982), Utterback and Abernathy (1975), Hill and Utterback (1979) and Utterback (1993) all represent attempts to identify characteristic patterns of the evolution of technology. Sahal starts from the rejection of either exclusively demand-pull theories of technological development or of pure supply-push theories, maintaining that technology both shapes its socioeconomic environment and is shaped by it. He stresses in particular the influence of scale and size on the evolution of technology: ultimately, the processes of scaling up (or of miniaturisation) reach limits and at this time radical innovations are needed to open up new 'avenues of innovation'. Some of those avenues may be so broad that they afford new opportunities in many sectors – recalling Nelson and Winter's 'generalised natural trajectories'.

Sahal's approach, like that of Metcalfe (1981) points to the importance of cyclical phenomena in the growth of industries and technologies and to the importance of *timing* in relation to public policy. Utterback and Abernathy (1975) and Abernathy and Clark (1985) also stress the cyclical path of evolution of technology and their work is of particular interest in relation to the issue of *process* innovations associated with scaling up. It also indicates an important link between theories of evolution of technology and the problems of firm strategies discussed by Teece *et al.* (1990), Teece (1987), Coombs and Richards (1991), Dodgson (1991) and many others. Neo-Schumpeterians have not on the whole succeeded in developing a behavioural theory relating firm *strategy* to routines and rules of thumb (but see Quintella, 1993; Mulder and Vergragt, 1990). Business studies theorists and economists have remained too far apart in this area.

There have been relatively few attempts to place the whole discussion of technological trajectories, patterns of innovation and the selection environment in the wider context of the evolution of the economic system as a whole. Many authors have followed Dosi (1982) in drawing a parallel between Kuhn's idea of 'paradigms' in the development of science and the evolution of technology. But the notion of a 'techno-economic paradigm', as put forward by Carlota Perez (1983, 1985, 1989) in her papers on microelectronics and world structural and institutional change, has several original distinguishing features.

In the first place, her concept of a change in the 'techno-economic paradigm' is one of a change in the basic approach and 'common-sense' of designers, engineers and managers which is so pervasive that it affects almost all industries and sectors of the economy. Secondly, she argues that the *economic* motivation for such a change of paradigm lies not only in the availability of a cluster of radical innovations offering numerous new potential applications, but also in the *universal* and *low cost* availability of a key factor or combination of factor inputs. This recalls Schumpeter's 'successive industrial revolutions' based on new factor combinations. Finally, she argues that before a new techno-economic paradigm can generate a new wave of world-wide economic growth, there is a period of adaptation of the socio-institutional framework, corresponding to the recession and depression phases of Schumpeter's 'long waves' of economic development. The old institutions suffer from inertia (see also Olson, 1982) and were in any case adapted to a now increasingly obsolete technological style and hence tend to 'lock out' alternative systems. There is therefore a period of 'mis-match' or structural adjustment between the new technology and the old framework affecting especially education and training, but also management structures, capital and labour markets, standards and regulation systems (see also Andersen, 1991). Perez therefore offers a link between the cyclical theories of technological evolution advanced by Sahal, Utterback, Abernathy and others and the theories of path-dependency, structural change and 'lock out' of alternatives put forward by Arthur, David, Dosi and others. The French Regulation School (Boyer, 1988) has many points of resemblance but originally did not attribute such a big role to technical change (see also the concept of 'socio-technical paradigms', Roobeek, 1987).

For Perez as for the 'Regulation School', the problem of stable growth is fundamentally a cyclical one. Periods of 'structural adjustment' are characterised by a high degree of instability and institutional turmoil but they are succeeded by quite long periods of prosperous growth based on the achievement of a good match between new technology, new routines and procedures in the sphere of the economy and other new institutions.

Thus, both at the micro- and the macro-level many neo-Schumpeterians look to institutional change to achieve and maintain harmonious growth and development.

A far more controversial twist to long-run theories of evolutionary development has come from Gomulka (1990). After many penetrating comments on the former East European national systems of innovation he comes by an ingenious route to the paradoxical conclusion that Schumpeter was right after all and the future will be a slow-down of innovation and a socialist system for reasons of social justice (see also the subtle analysis of Schumpeter's 'Capitalism, socialism and democracy' by Heertje, 1992).

From this it is evident that the neo-Schumpeterians as well as other institutional economists have breathed some new life into the debates which were central to classical political economy and have always been the pre-occupation of historians, but had been somewhat neglected by mainstream theory (Hodgson, 1993).

8 Conclusions

Clearly, it has not been possible within the limits of a relatively short survey article to cover an exploding field in all its depth and richness. As indicated in section 1, entire topics have been left aside, including some where neo-Schumpeterian research has made an outstanding contribution, such as international trade (e.g., Krugman, 1990; Soete, 1981, 1987; Dosi, Pavitt and Soete, 1990; Posner, 1961; Walker, 1977; Hufbauer, 1966; Vernon, 1966; De la Mothe, ed., 1990; Fagerberg, 1992; Dalum, 1992) and others where it has done very little, such as consumer behaviour. For the same reasons, this section concentrates only on a few of the most important topics.

Measurement

Neo-Schumpeterians have made original and ingenious use of a wide variety of statistics for the measurement of various aspects of technical change (scientometrics, technometrics, patents, R&D expenditures and personnel, innovations, diffusion rates and so forth). They have certainly not been uncritical in these endeavours and have initiated improvements in many areas (Grupp, 1991; Pavitt, 1985; Basberg, 1987; Griliches, 1990a; Irvine et al., 1987; Freeman, 1962; Freeman, ed., 1987; OECD, 1992b; Martin and Irvine, 1981, 1983, 1985; Soete and Wyatt, 1983; Soete et al., 1989; Archibugi et al., 1987; Archibugi and Pianta, 1992). They have been fortunate in the encouragement and support they have received from the OECD Directorate for Science, Technology and Industry, US National Science Foundation and UNESCO on Science and Technology Indicators. The OECD took the lead in standardisation of R&D statistics and has worked hard at improving them (OECD, 1963, 1970, 1976, 1981, 1993). It has also stimulated the development of new statistics intended to measure the 'output' of R&D (e.g., OECD, 1992b).

Nevertheless, as the neo-Schumpeterians certainly recognise, there are still some serious gaps and problems with the available statistics which have been pointed out in sections 4, 5 and 6. In particular, it is often unsatisfactory to use R&D expenditure statistics as a surrogate for all those activities at the level of the firm which are directed towards knowledge accumulation, technical change and innovation. We have measures of 'capital-intensity' and of 'energy-intensity' but not of 'knowledge-intensity'. There will always be problems in defining and measuring 'knowledge-intensity' but a more serious attempt will be needed in the 1990s and twenty-first century. Now that the role of 'intangible investment' has been generally recognised as equal in importance or even more important than fixed investment (World Bank, 1991; OECD, 1992a) it should be possible to make significant progress. These comments should not of course be taken as a denial of the great importance of R&D in the twentieth century. R&D remains at the heart of the network of institutions promoting technical change and is itself evolving (see, e.g., Meyer-Krahmer, 1990, 1992).

Another major gap where there has however already been some progress, is in the measurement of innovation outputs. The OECD (1992b) *STI Review* (December) reports the results of innovation measurement and survey work in six different European countries (Smith and Vidrei: Norway; Deiaco: Sweden; Kleinknecht and Reijnen: Netherlands; Cesaratto and Sirilli: Italy; Scholz: Germany; Auzeby and Francois: France). As the OECD (1992b) Introduction points out, these surveys are 'quite heterogeneous in terms of objectives, methods, definitions and so on' but nevertheless add considerably to our knowledge about innovation. Moreover, these efforts and earlier work have made it possible for the OECD to produce Guidelines for Standard Statistical Practice in this field (The 'Oslo Manual') and to work with the EC on a European Innovation Survey in 1993 based partly on the Manual.

This is a good example of fruitful interaction between academic research, national statistical offices in various countries, industrial agencies and firms and international organisations to generate valuable new data and analysis. It is to be hoped that it will be followed by much further work on scientific and technical services, intangible investment, diffusion of innovations and skills. However, there are still big problems with definitions (Winter, 1986b, 1987, 1988) classification and measurement to be resolved in all these areas.

Industry-specific studies

Neo-Schumpeterians have made a significant response to the plea of Nelson and Winter (1977) for greater attention to industry-specific features of technical change. Nevertheless, as noted in section 5, there are still many industries which have received very little attention. In particular, in common with industrial economics in general, far too little attention has been paid to the tertiary sector. Agriculture (e.g., Griliches, 1958; Ruttan, 1982; Feller *et al.*, 1987), mining (e.g., Townsend, 1976; Surrey, 1992) and especially manufacturing (see section 5) have all been fairly intensively studied, but service industries hardly at all.

This is very unsatisfactory for many reasons, principally, of course, the fact that service industries now account for nearly three-quarters of total employment in some industrial countries. A second reason is that they are now deeply affected by new technologies, in particular, information technology. For example, financial services are becoming very capital intensive and in particular, computer intensive; however, from the little work that has been done, it seems that they may not yet be very knowledge intensive. This is a rapidly changing situation and illustrates exactly the kind of transformation which merits the attention of neo-Schumpeterian research. Certainly there have been some studies (e.g., Heertje, ed., 1988; Baba and Takai, 1990; Barras, 1990; Cassiolato, 1992; Christensen, 1992) but they do not yet measure up to the scale of the problems. Organisational innovations in particular have been relatively neglected.

Perhaps even more important and an even bigger gap are the new and rapidly expanding services associated with computerisation: above all, software, multimedia, data banks and telematic services. Again, of course, there have been some pioneering studies (e.g., Thomas and Miles, 1989; Miles and Thomas, 1990; Miles, Schneider and Thomas, 1991; Mansell, 1988, 1989, 1990a; Quintas, ed., 1993; Brady and Quintas, 1991) but the rate of change is so rapid and the emerging structures, services and products so difficult to define and measure that they merit far more continuous study and research over a considerable period.

Networks, concentration and firm behaviour

As indicated in sections 3 and 7, there are major theoretical issues which are still unresolved in the exciting area of firm behaviour, interfirm networking and so-called 'globalisation'. The neo-Schumpeterians have surely been right to emphasise the evolutionary and changing aspects of firm organisation and behaviour (Chandler, 1977, 1990; Winter, 1964, 1971, 1986b, 1988; Witt, ed., 1993). Their vision of the firm as a learning, innovating organisation, their insistence on the heterogeneity of firms (Dosi, 1984; Nelson, 1991, 1992b) and their observations of the multiple external knowledge and information linkages of firms have brought realism and plausibility to economic theory in an area which was in danger of losing touch altogether with the realities of everyday life. Schumpeter was right to exonerate Walras and Marshall from the simplistic charge of believing that the entrepreneur could have omniscient knowledge (Andersen, 1992b) but both Schumpeter and neo-Schumpeterians are surely right in their critique of the poverty-stricken assumptions that price signals contained sufficient information to explain the most important features of firm behaviour, or that firms behave 'as if' they had perfect knowledge and foresight.

In particular, neo-Schumpeterians have given an extremely interesting account of Japanese firms. However, it is by no means clear whether the 'J-firm' is culturally specific to Japan, whether the activities of Japanese MNCs will diffuse its essential features, through the world, or indeed, whether Japanese firms will adopt some features of their American and

European competitors. In some senses, the 'networking' firm has always been with us (Allen, 1983; Freeman, 1991a). The neo-Schumpeterians have not yet really clarified just what is new about the networking firms of the late twentieth century. Is it their use of ICT? – or their adoption of management techniques characteristic of J-Firms? Or is it a temporary upsurge of networking to cope with the need for technological complementarities associated with the new generic technologies? How do networking firms affect the theory of monopoly and oligopoly? It is very much to the credit of neo-Schumpeterians that they are alive to new trends in firm behaviour and some of the effects of diffusion of new technologies and it would also be unfair to expect them to provide complete answers to some questions which puzzle everyone. Nevertheless, these questions indicate that the neo-Schumpeterians still have a long way to go in their theoretical analysis as well as in their empirical research on such questions as the behaviour of Japanese firms outside Japan and the behaviour of American and European subsidiaries inside Japan, and the evolution of networks over time. Indeed for people who believe above all in evolutionary development, the approach of some neo-Schumpeterians to networks has been surprisingly 'static' (but see Lundgren (1991) for an extremely interesting longitudinal analysis of image processing networks in Sweden).

Models

In view of their emphasis on diversity of firms, industries and economic systems, as well as discontinuities in technical change, and their reluctance to accept easy simplifying generalisations too remote from reality, it is perhaps surprising that neo-Schumpeterians have in fact already generated quite a large family of interesting models. In his review of this new generation of models, however, Silverberg (1988) points out that there were early predecessors, such as Goodwin's (1951) demonstration that self-sustaining business cycles were only possible in the context of non-linear models. Boyer (1993) also points to the importance of Keynesian and Kaldorian models in the development of evolutionary approaches (see also Boyer and Petit, 1989).

As Boyer (1988, 1993) shows, the findings of neo-Schumpeterian research (sections 3–7 above) have had two rather different outcomes. On the one hand, they have led some economists to try and improve general equilibrium modelling by extending it to include more realistic assumptions about the externalities associated with innovation, networks and education and increasing returns. This is the direction of the so-called 'new growth theory' (Grossman and Helpman, 1991; Romer, 1986, 1991; Aghion and Howitt, 1993). On the other hand, most neo-Schumpeterians have preferred to follow the pioneering approach of Nelson and Winter (1974, 1982) and to discard altogether the general equilibrium approach in favour of evolutionary modelling (e.g., Silverberg, 1984, 1987, 1988, 1990; Smith, 1991; Silverberg *et al.*, 1988; Silverberg, Dosi and Orsenigo, 1988; Silverberg and Lehnert, 1992; Amable, 1992, 1993a; Iwai, 1984a, 1984b; Gibbons and Metcalfe, 1986; Chiaromonte and Dosi, 1993; Dosi *et al.*, 1992; Weidlich and Braun, 1992; and see also Arthur *et al.*, 1987, and the references to Arthur and David in section 2).

These and other simulation models have certainly demonstrated the feasibility of modelling aspects of economic behaviour which were previously ignored (innovations and their diffusion, various appropriability regimes, heterogeneity of agents, learning by interacting, quality improvements, dynamic efficiency and so forth). However, the diversity of agents, industries and national circumstances, the complexity of their interactions, and the lack of sufficient understanding of some of the key relationships, mean that these evolutionary models, although more realistic and richer in many respects than their predecessors, need to be constantly complemented and tested against the type of historical and empirical research which has been the hallmark of the neo-Schumpeterian tradition. Such research is in any case always essential since the reality which they attempt to represent is in a constant state of flux and a unique historical process. Probably the best example of such an evolutionary micro–macro model is that developed by Eliasson at IUI in Sweden (1986, 1990, 1991a, 1991b, 1991c). This work is particularly good in illustrating the importance of new entry and diversity for stable and rapid economic growth.

It is very much to be hoped that the neo-Schumpeterians above all do not fall into the trap of recent generations of economists in sacrificing descriptive realism and richness of historical evidence to the requirements of formal mathematisation. With this reservation, they have given a valuable new impulse to the prolonged efforts of the profession to provide satisfactory formal representation of the complex, untidy and changing behaviour of the economic system.

A new form of assessment?

This section has raised some criticisms of neo-Schumpeterian research as well as a rather positive general view of its achievements. The empirical research discussed in sections 3 – 7 has been a relatively strong feature of the neo-Schumpeterian tradition. But also on any showing, such books and papers as Nelson and Winter (1974, 1982); Nelson (ed., 1993); Rosenberg (1976, 1982, 1994), Stoneman (1983), Dosi *et al.* (1988), Antonelli, Petit and Tahar (1992), Aoki (1988), Arthur (1986, 1988, 1989), David (1976,

1985, 1991, 1993), Hodgson (1993), Lundvall (ed., 1992), Winter (1971, 1988), Dosi, Pavitt and Soete (1990) and Soete (1981, 1991) represent a significant contribution to economic theory, which challenges a number of cherished conventional ideas and assumptions. Clearly this is a biassed survey coming from someone who has been rather heavily involved in this type of research for nearly 35 years. It would be desirable therefore to invite a more critical survey from someone who has been closer to the mainstream of economic theory. But even better might be to innovate with a new form of assessment. Most refereeing work in any discipline and most literature survey work is done by specialists *within* that discipline or sub-discipline for obvious reasons. However, there would be a good argument in this case for inviting an assessment or critical survey from well-qualified people *outside* the discipline of economics, i.e., an international panel of engineers, biologists, physicists, historians, geographers, sociologists, political scientists, psychologists and scholars from business studies.

An assessment from neighbouring disciplines is desirable in any case for several reasons, especially:

(i) The overlap with several of these disciplines is considerable and in some cases inter-disciplinary research is already important. Collaboration with engineers and scientists is in any case essential in this field;

(ii) There is a possibility of a more objective and certainly more independent view of some fundamental issues than from within the economics discipline.

A century ago the 'Methodenstreit' divided the German economics profession and raised many similar questions. Whilst the neo-Schumpeterians would probably not wish to be identified with Schmoller or the neo-classicals necessarily with Carl Menger, the basic questions of how much does history matter, the role of country-specific institutions, and the limits of universal generalisations about economic behaviour are still with us. Schumpeter attempted (and according to most accounts failed) to bridge the warring camps in the Methodenstreit. It could be a fruitful experiment, although difficult to set up, to see if an interdisciplinary assessment could make any contribution to the dilemmas of the economics profession today.

3 Uneven (and divergent) technological accumulation among advanced countries: evidence and a framework of explanation

PARIMAL PATEL AND KEITH PAVITT

1 The persistence of technology gaps

The renewed interest over the past ten years in the nature and determinants of international patterns of economic growth has confirmed that international 'catch up' in technology and productivity is neither automatic nor easy, since it depends on investment in tangible capital, and intangible capital in the form of education and training and – at least in the industrially advanced countries – of business expenditures on R&D and related activities (Fagerberg, 1987, 1992). These factors explain why some developing countries have been successful in reducing the technology and productivity gap, whilst others have not.

This is because the international diffusion of technology is neither automatic nor easy (see Bell and Pavitt, 1993). Both material artefacts and the knowledge to develop and operate them are complex, involving multiple dimensions and constraints that cannot be reduced entirely to codified knowledge, whether in the form of operating instructions, or a predictive model and theory. Tacit knowledge – underlying the ability to cope with complexity – is acquired essentially through experience, and trial and error. It is misleading to assume that such trial and error is either random, or a purely costless by-product of other activities like 'learning by doing' or 'learning by using'. Tacit (and other forms of) knowledge are increasingly acquired within firms through deliberately planned and funded activities in the form of product design, production engineering, quality control, education and staff training, research, or the development and testing of prototypes and pilot plant. Differences amongst countries in the resources devoted to such deliberate learning – or 'technological accumulation' – have led to international technological gaps which, in turn, have led to international differences in economic performance.

But while uneven and divergent development is readily acknowledged amongst the *developing* countries, the same is not true for the *advanced* (OECD) countries. Until recently, it was commonly assumed that the open trading system would allow the rapid international diffusion of technology, so that the catching up of Western Europe and Japan to the levels of technology and efficiency of the world's leading country (the USA) would be relatively smooth. In fact, there has also been uneven development amongst industrially advanced countries. Some (e.g., the UK; see Pavitt, 1980) have caught up only very partially, whilst others (e.g., FR Germany and Japan) have actually overtaken the world's technological leading country – the USA – in certain important sectors (Nelson, 1990b). At a more aggregate level, Soete and Verspagen (1993) have recently shown that productivity convergence in the OECD countries stopped at the end of the 1970s.

Thus, technology gaps amongst the industrial countries have not been eliminated. Hence the continuing relevance of the 'neo-technology' theories of trade and growth, that were pioneered by Posner (1961) and Vernon (1966), and confirmed by Soete (1981) and Fagerberg (1987, 1988a), as well as by the company-based analyses of Cantwell (1989), Franko (1989) and Geroski *et al.* (1993). Hence, also, the growing interest in the implications of international technology gaps for policy (Ergas, 1984) and for theory (Dosi *et al.*, 1990).

We shall now present statistical evidence of uneven and divergent technological accumulation in the 1980s, and shall argue that technological gaps amongst the OECD countries will not be eliminated in the 1990s. Given that the activities contributing to technological accumulation are complex and varied, all statistical measures are bound to be imperfect. However, as a result of the growing demands from public and private policy makers for better data, progress has been made in both measurement and conceptualisation. The advantages and drawbacks of the various measures have been extensively reviewed elsewhere (Freeman, 1987; Griliches, 1990a; Patel and Pavitt, 1995). In particular, we have shown in our earlier work that the combined use of data on R&D activities, and on patenting in the USA by country of origin, gives a plausible and consistent picture of technological activities at the world's technological frontier.[1]

2 The evidence of uneven (and divergent) technological accumulation

2.1 Amongst countries in the volume of technological activities

The data on R&D and US patenting activities show no evidence of convergence in national capacities for technological accumulation since the early 1970s, and some evidence of divergence in the 1980s.

Table 3.1 presents trends in the percentage of Gross Domestic Product spent by business on R&D activities in 17 OECD countries since 1967.[2] These show a certain stability in the rankings throughout the period at the two ends of the spectrum: Switzerland has remained with the highest share, and Ireland, Spain and Portugal with the three lowest shares. Otherwise there are countries who started near the top but have moved down the rankings: Canada, the Netherlands and – above all – the UK; there are also countries that have improved their positions: FR Germany, Sweden, Japan and – above all – Finland. In general, stability in the rankings of the countries is confirmed by a statistically significant (positive) correlation between their ranks in 1967 and in 1991.[3]

Overall, there are no statistical signs of convergence in the industry-funded shares over time, since the standard deviation of the distribution has not decreased over time. On the contrary, it has increased markedly in the 1980s, suggesting technological divergence amongst countries. In this context, it is worth noting that the US share began slipping progressively below that of FR Germany, Japan, Sweden and Switzerland in the 1970s, and that the gap grew much larger in the 1980s.

Table 3.2 shows trends in per-capita national patenting in the USA for the same 17 OECD countries. At first sight the evidence about divergence is more ambiguous. When the USA is included, the standard deviation of the population increases between the late 1960s and the early 1970s – thereby suggesting divergence – but then decreases until the mid 1980s, after which it increases again to its original level. However, there are well-known reasons for excluding the USA from such a comparison, since for firms in the USA, we are measuring domestic patenting, whereas for firms in other countries we are measuring foreign patenting. Given the propensity of firms to seek patent protection more intensely in their home country (Bertin and Wyatt, 1988), the rate of technological accumulation in the USA is overestimated. At the same time, given the tendency of firms to give increasing attention to patenting in foreign markets, trends over time will tend to overestimate any decline in US performance (Kitti and Schiffel, 1978), and thereby show a spurious degree of convergence.

When the USA is excluded, the evidence in table 3.2 on the whole confirms that of table 3.1. Throughout the period, Switzerland stays at the top, and Ireland, Spain and Portugal at the bottom. Britain's relative position declines, whilst Finland, FR Germany and Japan improve. In general, stability in the rankings of the countries is confirmed by a significant and positive correlation between their ranks in 1963–8 and 1986–90.[4] At the same time there is an indication of international divergence in that the standard deviation increases over the period. The one anomaly is the reduction in the standard deviation in 1981–5, but this may reflect the reduction in the

Table 3.1 Trends in industry financed R&D as a percentage of GDP in 17 OECD countries; 1967–1991

	1967	1969	1971	1975	1977	1979	1981	1983	1985	1987	1989	1991
Belgium	0.66	0.64	0.71	0.84	0.91	0.95	0.96	1.02	1.09	1.16	1.14	1.16
Canada	0.40	0.39	0.38	0.33	0.32	0.39	0.49	0.45	0.56	0.57	0.54	0.59
Denmark	0.34	0.39	0.41	0.41	0.41	0.42	0.46	0.53	0.60	0.66	0.71	0.85
Finland	0.30	0.32	0.44	0.44	0.49	0.53	0.62	0.72	0.89	0.98	1.07	1.07
France	0.60	0.64	0.67	0.68	0.69	0.75	0.79	0.88	0.92	0.92	0.98	0.99
FR Germany	0.94	1.03	1.13	1.11	1.12	1.32	1.40	1.48	1.65	1.80	1.78	1.57
Ireland	0.19	0.23	0.30	0.23	0.22	0.23	0.26	0.27	0.35	0.40	0.45	0.58
Italy	0.33	0.38	0.44	0.43	0.37	0.40	0.43	0.42	0.49	0.49	0.56	0.61
Japan	0.83	1.00	1.09	1.12	1.11	1.19	1.38	1.59	1.81	1.82	2.05	2.13
Netherlands	1.12	1.04	1.02	0.97	0.87	0.86	0.83	0.89	0.96	1.11	1.07	0.91
Norway	0.35	0.39	0.41	0.49	0.49	0.50	0.50	0.61	0.80	0.88	0.81	0.77
Portugal	0.04	0.06	0.09	0.05	0.04	0.09	0.10	0.11	0.11	0.11	0.14	0.14
Spain	0.08	0.08	0.11	0.18	0.18	0.18	0.18	0.22	0.25	0.29	0.34	0.38
Sweden	0.71	0.69	0.80	0.96	1.07	1.11	1.24	1.45	1.71	1.74	1.68	1.71
Switzerland	1.78	1.78	1.67	1.67	1.71	1.74	1.68	1.67	2.16	2.13	2.07	2.07
United Kingdom	1.00	0.92	0.81	0.80	0.80	0.82	0.91	0.86	0.95	1.02	1.04	0.94
United States	0.99	1.03	0.97	0.98	0.98	1.05	1.17	1.31	1.42	1.37	1.36	1.36
Standard deviation												
All countries	0.46	0.46	0.43	0.43	0.45	0.47	0.48	0.52	0.61	0.61	0.60	0.58
Excluding the US	0.47	0.46	0.43	0.44	0.45	0.47	0.48	0.52	0.62	0.62	0.62	0.59

Source: OECD.

Table 3.2 *Trends in per-capita patenting in the US from 17 OECD countries*
(US patents per million population)

	1963-8	1969-74	1975-80	1981-5	1986-90
Switzerland	138.0	197.1	207.1	179.2	193.6
United States	236.6	244.7	181.3	156.7	177.9
Japan	9.9	38.9	56.0	82.5	139.4
FR Germany	54.7	86.6	91.8	97.7	122.2
Sweden	64.2	94.7	100.9	87.1	99.6
Canada	41.3	55.8	49.0	45.8	63.1
Netherlands	35.9	48.8	46.8	47.0	60.4
Finland	5.1	14.8	22.3	30.8	50.4
France	26.1	41.0	39.7	38.8	49.6
United Kingdom	43.3	55.8	46.6	40.1	47.8
Denmark	18.4	31.9	29.7	27.9	35.7
Belgium	16.3	29.3	26.4	23.8	30.4
Norway	12.7	20.3	23.0	19.4	27.2
Italy	7.8	13.1	13.0	14.0	20.2
Ireland	2.0	6.5	5.1	6.7	12.9
Spain	1.2	2.1	2.3	1.6	3.1
Portugal	0.3	0.6	0.3	0.4	0.6
Standard deviation					
All countries	60.56	67.56	59.44	52.05	59.26
Excluding the US	35.05	48.91	51.44	46.13	53.58

Source: Based on data supplied to SPRU by the US Patent and Trademark Office.

overall number of patents granted, following a reduction in the number of patent examiners (see Griliches, 1990a).

Finally, table 3.3 presents recent trends in patenting in the USA by a number of developing countries. We are very much aware of the inadequacies of US patenting as a measure of the largely imitative activities in technological accumulation that are performed in developing countries. Studies using other approaches have shown the superior performance of East Asian countries, compared with those of Latin America and with India (see, for example, Dahlman *et al.*, 1987). Table 3.3 simply shows that, whilst most of the developing countries have continued with a very low level of US patenting, Taiwan and S. Korea have both seen massive increases. This indicates that technology in Taiwan and S. Korea is now attaining world best practice levels in an increasing number of fields – a striking example of technological catch-up compared with the advanced countries,[5] and of technological divergence compared with other developing countries.

Table 3.3 US patenting activities of selected developing countries, 1969–1992

Country	1969	1970	1971	1972	1973	1974	1975	1976	1977	1978	1979	1980	1981	1982	1983	1984	1985	1986	1987	1988	1989	1990	1991	1992
Taiwan	0	0	0	1	0	0	23	28	52	29	38	65	79	88	65	97	174	208	343	457	592	732	904	1000
South Korea	0	3	2	7	5	7	11	7	5	12	4	8	15	14	26	29	38	45	84	97	159	225	402	538
China P.Rep.	5	6	15	8	10	22	1	6	1	0	2	1	3	0	1	2	1	9	23	47	52	47	52	41
Hong Kong	7	8	19	7	15	9	10	20	9	21	13	27	33	18	14	24	25	30	34	41	48	52	50	60
Mexico	67	43	63	43	42	51	66	78	42	24	36	41	43	35	32	42	32	37	49	44	39	32	28	39
Brazil	18	17	14	16	18	21	17	18	21	24	19	24	23	27	19	20	30	27	34	29	36	41	61	40
Venezuela	6	3	13	7	5	7	0	0	0	2	11	11	12	10	5	11	15	21	24	20	23	n.a.	n.a.	n.a.
Argentina	17	23	22	29	27	24	24	20	21	24	18	25	18	21	20	11	17	18	16	20	17	16	n.a.	20
Singapore	2	0	4	4	7	6	1	3	3	2	0	3	4	3	5	4	9	3	11	6	18	17	n.a.	20
India	18	16	10	19	21	17	13	17	14	14	6	4	4	14	12	10	18	12	14	14	18	23	22	24

Source: Based on data supplied to SPRU by the US Patent and Trademark Office.

Table 3.4 *Vocational qualifications of the workforce in Britain, the Netherlands, Germany, France and Switzerland*

Level of qualification	Britain 1988	Netherlands 1989	Germany 1987	France 1988	Switzerland 1991
University degrees	10	8	11	7	11
Higher technician diplomas	7	19	7	7	9
Craft/lower tech. dips.	20	38	56	33	57
No vocational qualifications	63	35	26	53	23
Total	100	100	100	100	100

Source: Prais, 1993.

2.2 Amongst countries in work-force education and training

International comparisons of education and training over the past ten years have moved well beyond the average numbers of years of schooling that used to be common usage. Greater attention is now paid to the distribution of education levels amongst different groups in the working population, and to quality as measured through educational attainment. One of the main pioneers has been S. Prais (together with his colleagues) at the National Institute of Economic and Social Research in London.[6] Some of the major results of their work are summarised in table 3.4, which uses census data to compare the vocational qualifications of the workforce in five European countries.

It shows striking *similarities* between countries in the proportion with university degrees (7–11 per cent), but even more striking *differences* in the proportions with intermediate qualifications (66 per cent in Switzerland – 27 per cent in the UK), and with no vocational qualifications (63 per cent in the UK – 23 per cent in Switzerland). Although there had been some improvement in the UK position in the 1980s, the qualifications gap between Germany and France, on the one hand, and the UK, on the other, actually widened (Patel and Pavitt, 1991b). These skill levels in the workforce are reflected in productivity differences resulting from differences in machine maintenance, consistency in product quality, workforce flexibility, and learning times on new jobs.

These studies have been supplemented by comparisons of educational attainment across countries, and which tend to confirm their findings: thus, Dutch adolescents are two to three years ahead of their English counterparts in mathematical attainment (Mason *et al.*, 1992). Over a broader geographical area, similar differences are found, with adolescents from Japan, the four East Asian 'tiger' countries, and continental Europe (including

Hungary) clearly out-performing their counterparts from the USA (and often the UK) in mathematics (Newton *et al.*, 1992).

2.3 Among countries in the sectoral composition of national technological activities

So far, we have compared countries' aggregate technological performance. Table 3.5 shows the sectoral patterns of technological advantage of 19 OECD countries. On the basis of the US patent classification, technologies have been divided into 11 fields. The content of most of them will be clear from their titles: technologies for extracting and processing raw materials are related mainly to food, oil and gas; defence-related technologies are defined as aerospace and munitions. For each country-region and technological field, we have calculated an index of 'Revealed Technology Advantage' (RTA) in 1963–8 and 1985–90.[7]

Table 3.5 shows markedly different patterns and trends amongst the three main, technology-producing regions of the world – USA, Europe and Japan – in their fields of technological advantage and disadvantage. The USA has seen rapid decline in motor vehicles and consumer electronics; growing relative strength in technologies related to weapons, raw materials and telecommunications; and an improving position in chemicals. In Japan, almost the opposite has happened: growing relative strength in electronic consumer and capital goods and motor vehicles, together with rapid relative decline in chemicals, and continued weakness in raw materials and weapons. In Western Europe, the pattern is different again, and very close to that of its dominant country – FR Germany: continuing strength in chemicals, growing strength in weapons, continued though declining strength in motor vehicles, and weakness in electronics.

Table 3.6 examines the similarities and differences amongst countries' technological specialisations in greater and more systematic detail.[8] It uses correlation analysis to measure both the stability over time of each country's sectoral strengths and weaknesses in technology (first row), and the degree to which they are similar to those of other countries (correlation matrix). The first row shows that, with five exceptions (Australia, Ireland, Italy, Portugal and the UK), most OECD countries have a statistically significant degree of stability in their technological strengths and weaknesses between the 1960s and the 1980s: ten at the 1 per cent level, and a further four at the 5 per cent level, thereby confirming the path-dependent nature of national patterns of accumulation of technological knowledge.

The correlation matrix also confirms the differentiated nature of technological knowledge, with the very different strengths and weaknesses in Japan, the USA and Western Europe: each is negatively correlated with the

Table 3.5 Sectoral patterns of revealed technological advantage*, 1963–1968 to 1985–1990

		Fine chemicals	Industrial chemicals	Materials	Mechanical Engineering	Vehicles	Electrical machinery	Electronic cap. goods	Telecomms.	Electronic Consumer goods	Raw material related	Defence related
USA	1963-8	0.89	0.93	1.04	1.01	0.89	1.00	1.02	1.03	0.94	1.08	0.99
	1985-90	0.97	0.98	0.95	0.99	0.55	1.01	0.97	1.04	0.65	1.28	1.15
Europe**	1963-8	1.34	1.29	0.86	0.99	1.48	1.00	0.92	0.91	1.26	0.61	1.14
	1985-90	1.33	1.19	0.83	1.13	1.02	0.92	0.61	0.94	0.59	0.83	1.40
Japan	1963-8	2.95	1.62	1.02	0.77	0.83	1.17	1.47	1.06	1.99	0.44	0.36
	1985-90	0.72	0.92	1.42	0.85	2.21	1.08	1.65	0.97	2.50	0.37	0.09
Australia	1963-8	1.05	0.69	0.80	1.16	1.44	0.74	0.19	0.72	1.48	1.02	0.31
	1985-90	0.80	0.65	0.42	1.21	1.35	0.59	0.30	0.54	0.34	1.82	1.63
Austria	1963-8	1.41	0.80	1.13	1.25	1.21	0.62	0.28	0.38	1.69	0.39	0.26
	1985-90	0.84	0.73	0.68	1.39	1.94	0.82	0.32	0.43	0.41	0.90	1.96
Belgium	1963-8	1.23	1.38	3.99	0.71	0.44	0.92	0.69	1.02	3.97	0.42	0.78
	1985-90	1.85	1.79	2.21	0.77	0.19	0.98	0.21	0.50	1.24	1.10	0.80
Canada	1963-8	0.71	0.81	0.75	1.11	1.35	0.78	0.56	0.85	0.36	1.43	0.71
	1985-90	0.68	0.74	0.67	1.15	0.65	0.83	0.40	1.38	0.47	1.69	1.13
Denmark	1963-8	3.05	0.77	0.97	1.11	0.39	0.87	0.47	0.59	1.17	0.92	0.12
	1985-90	2.38	0.91	0.49	1.19	0.20	0.90	0.21	0.46	0.54	0.82	0.39
Finland	1963-8	0.00	0.62	0.00	1.30	2.17	0.72	0.00	0.31	0.20	1.45	0.47
	1985-90	0.88	0.76	0.76	1.54	0.71	0.68	0.10	0.51	0.15	1.13	0.82
France	1963-8	1.86	1.02	1.05	1.02	2.11	1.23	0.84	1.15	0.81	0.49	1.11
	1985-90	1.34	1.03	0.88	1.04	0.57	1.17	0.85	1.56	0.49	1.03	1.55
Germany	1963-8	1.12	1.49	0.68	0.96	1.43	0.78	0.92	0.74	1.88	0.54	1.04
	1985-90	1.14	1.37	0.85	1.18	1.42	0.87	0.52	0.79	0.54	0.61	1.58

Table 3.5 (*cont.*)

		Fine chemicals	Industrial chemicals	Materials	Mechanical Engineering	Vehicles	Electrical machinery	Electronic cap. goods	Telecomms.	Electronic Consumer goods	Raw material related	Defence related
Ireland	1963-8	0.00	0.39	0.00	1.29	3.00	0.00	0.00	0.65	0.00	0.55	0.00
	1985-90	1.57	1.18	0.55	0.87	0.63	1.38	0.94	0.72	0.99	1.66	0.00
Italy	1963-8	1.29	1.93	0.51	0.93	1.26	0.68	0.87	0.69	0.53	0.71	0.78
	1985-90	1.77	1.10	0.62	1.15	1.21	0.73	0.76	0.85	0.41	0.78	0.92
Netherlands	1963-8	1.71	1.46	1.24	0.75	0.15	1.34	1.90	1.11	1.95	1.15	0.15
	1985-90	0.54	1.13	0.89	0.87	0.26	1.27	1.24	1.25	1.82	1.07	0.33
Norway	1963-8	0.94	0.63	0.00	1.25	0.36	1.16	0.65	0.47	0.29	0.91	0.46
	1985-90	0.83	0.62	0.27	1.14	0.33	0.70	0.24	0.92	0.43	2.30	1.79
Portugal	1963-8	10.58	1.41	0.00	0.99	0.00	0.67	0.00	0.00	0.00	0.98	3.46
	1985-90	1.90	1.66	0.00	1.11	0.00	0.43	0.00	0.00	0.00	2.97	0.00
Spain	1963-8	0.84	0.56	0.48	1.20	3.00	0.86	0.13	1.02	0.47	0.47	1.93
	1985-90	1.93	0.76	0.37	1.22	1.84	0.75	0.12	0.55	0.07	0.85	2.72
Sweden	1963-8	0.94	0.44	0.44	1.22	1.16	1.14	0.72	1.37	0.38	0.68	2.35
	1985-90	0.72	0.57	0.58	1.43	0.93	0.89	0.33	0.79	0.23	0.99	1.86
Switzerland	1963-8	2.60	2.01	0.27	0.90	0.56	0.85	0.56	0.73	0.63	0.47	1.45
	1985-90	1.81	1.48	0.63	1.12	0.51	0.78	0.41	0.64	0.43	0.72	1.16
UK	1963-8	0.88	1.03	1.09	1.02	1.99	1.22	1.04	1.03	0.93	0.70	1.28
	1985-90	1.83	1.05	0.92	1.04	0.86	0.88	0.71	1.07	0.73	0.99	1.37

Notes:

* For the definition of the Revealed Technology Advantage Index see footnote 7 in the text.

** Europe is defined as the 15 European countries included in this table.

Source: Based on data supplied to SPRU by the US Patent and Trademark Office.

Table 3.6. *Stability and similarities amongst countries in their sectoral specialisations: correlations of revealed technology advantage indices across 34 sectors*

	Australia	Austria	Belgium	Canada	Denmark	Finland	France	Germany	Ireland	Italy	Japan	Neth'land	Norway	Portugal	Spain	Sweden	Switl'and	UK	USA
Stability: correlations over time, 1963-8 to 1985-90																			
	0.28	0.76*	0.54*	0.67*	0.47*	0.59*	0.82*	0.35*	0.05	0.32	0.45*	0.66*	0.35*	0.25	0.53*	0.73*	0.83*	0.23	0.55*
Similarities: correlations amongst countries, 1985-90																			
Austria	0.36*																		
Belgium	-0.09	-0.14																	
Canada	0.52*	0.47*	0.05																
Denmark	0.18	-0.03	0.33*	0.32															
Finland	0.47*	0.45*	0.20	0.54*	0.45*														
France	-0.27	-0.16	0.10	-0.14	0.10	-0.15													
Germany	0.27	0.05	0.22	-0.18	0.21	0.32	0.29												
Ireland	0.07	-0.10	0.09	0.21	0.14	0.03	-0.09	-0.31											
Italy	0.28	0.28	0.06	0.34*	0.30	0.53*	-0.23	0.22	0.28										
Japan	-0.43*	-0.07	0.06	-0.44*	-0.22	-0.26	-0.44*	-0.20	-0.13	-0.13									
Netherlands	-0.24	-0.18	-0.03	0.06	-0.04	0.07	-0.33*	-0.38*	0.27	0.08	0.24								
Norway	0.36*	0.36*	-0.20	0.62*	0.22	0.28	0.02	-0.12	0.02	-0.03	-0.50*	-0.23							
Portugal	0.32	0.48*	0.17	0.31	0.11	0.43*	-0.09	0.15	0.11	0.35*	-0.20	-0.04	0.06						
Spain	0.32	0.13	-0.11	0.34*	0.68*	0.38*	0.00	0.28	-0.07	0.38*	-0.23	-0.28	0.41*	0.20					
Sweden	0.25	0.46*	-0.05	0.38*	0.40*	0.53*	0.36*	0.26	-0.07	0.19	-0.38*	-0.35*	0.30	0.30	0.45*				
Switzerland	0.35*	-0.12	0.11	-0.21	0.01	0.07	0.06	0.72*	-0.12	0.17	-0.19	-0.26	-0.14	0.04	0.13	-0.04			
UK	0.08	-0.15	-0.04	-0.10	0.40*	0.03	0.23	0.30	-0.02	-0.03	-0.34*	-0.20	0.17	0.00	0.32	0.10	0.20		
USA	0.22	-0.03	-0.20	0.42*	-0.02	-0.01	0.23	-0.37*	0.25	-0.09	-0.81*	-0.03	0.50*	0.06	-0.02	0.13	-0.26	0.11	
W. Europe	0.26	0.04	0.21	-0.08	0.33*	0.34*	0.49*	0.93*	-0.19	0.27	-0.41*	-0.37*	-0.02	0.17	0.35*	0.39*	0.73*	0.45*	-0.19

Notes:

For the definition of the Revealed Technology Advantage Index see footnote 7 in the text.

Europe is defined as the 15 European countries included in this table.

* Denotes Correlation Coefficient significantly different from zero at the 5 per cent level.

Source: Based on data supplied to SPRU by the US Patent and Trademark Office.

other two, and significantly so in two cases out of three (the USA with the other two regions). More generally, it confirms that countries tend to differ markedly in their patterns of technological specialisation.[9] Of the 171 correlations amongst pairs of countries in table 3.6, only 31 (18 per cent) are positively and significantly correlated at the 5 per cent level. Amongst these we find FR Germany similar to Switzerland (chemicals and machinery), and Canada similar to Australia, Finland and Norway (raw material-based technologies). Japan has a unique pattern of specialisation, with no significant positive correlations with other countries, but plenty of negative ones.

2.4 Implications

The above comparisons show some striking differences – even divergences – in the rate and direction of technological accumulation in the industrial countries. Those related to fields of technological specialisation reflect inevitable diversity in stages of economic and technological development, or a desirable diversity in fields of national scientific and technological specialisation. Those related to differences in the overall volume of investment in technological accumulation should be causes of disquiet because – if allowed to persist – they are likely to re-inforce any uneven and divergent rates of national technological and economic development in future.

The most striking international differences are in the volume of change-generating activities (including R&D) supported by business firms, and in the skills of the work force that they employ and that they (unequally) train.[10] Japan and FR Germany are the major countries with high company R&D and workforce skills, with the UK (and probably the USA) amongst the major industrial countries at the other end of the spectrum.

3 The effects of 'global corporations'

3.1 The importance of home countries for technological activities

The unfettered behaviour of large corporations is unlikely – in and of itself – to smooth out these international differences in technological accumulation, for three sets of reasons.

First, large firms do not play a major role in the development and control of some significant fields of technology: in particular, capital goods, components, and measuring and control instruments (Patel and Pavitt, 1991a). As the recent experience of the Japanese automobile industry shows, large firms can stimulate technological accumulation in these fields through their supplier networks, including those outside Japan. But this depends on the

Table 3.7 *Geographic location of large firms' US patenting activities, according to nationality; 1985–1990*
(Percentage shares)

Firms' nationality	Home	Abroad	Of which USA	Of which Europe	Of which Japan	Of which Other
Japan (143)	98.9	1.1	0.8	0.3	–	0.0
USA (249)	92.2	7.8	–	6.0	0.5	1.3
Italy (7)	88.1	11.9	5.4	6.2	0.0	0.3
France (26)	86.6	13.4	5.1	7.5	0.3	0.5
Germany (43)	84.7	15.3	10.3	3.8	0.4	0.7
Finland (7)	81.7	18.3	1.9	11.4	0.0	4.9
Norway (3)	68.1	31.9	12.6	19.3	0.0	0.0
Canada (17)	66.8	33.2	25.2	7.3	0.3	0.5
Sweden (13)	60.7	39.3	12.5	25.8	0.2	0.8
UK (56)	54.9	45.1	35.4	6.7	0.2	2.7
Switzerland (10)	53.0	47.0	19.7	26.1	0.6	0.5
Netherlands (9)	42.1	57.9	26.2	30.5	0.5	0.6
Belgium (4)	36.4	63.6	23.8	39.3	0.0	0.6
All firms (587)	89.0	11.0	4.1	5.6	0.3	0.9

Note:
The parenthesis contains the number of firms based in each country.
Source: Based on data supplied to SPRU by the US Patent and Trademark Office.

autonomous development of skills and technological capabilities amongst the suppliers themselves.

Second, technology-generating activities remain among the most domesticated of all corporate activities.[11] Table 3.7 shows that, in the second half of the 1980s, 89 per cent of the technological activities of the world's largest firms continued to be performed in their home country – a 1 per cent increase over the previous five-year period. Not unsurprisingly, the share performed in foreign countries by large firms based in smaller countries tends to be higher, although the proportion for Finnish firms (18.3 per cent) compared with that for British firms (45.1 per cent) shows that factors other than size are at work.

In particular, Cantwell (1992) has shown that the share of foreign firms in total production is the most important factor explaining differences amongst firms in the location of their technological activities, but that foreign patenting shares are smaller than foreign production shares. In other words, the technology intensity of foreign production is consistently and significantly less than that of home production, and concerned mainly with product and process adaptation to local conditions.

Third, table 3.7 also shows that the foreign technological activities of large firms are not globalised, but concentrated almost exclusively in the 'triad' countries – especially the USA and Europe (and, more specifically, Germany). More detailed data show that the largest proportionate increases in foreign technological activities in the 1980s were in British and Swedish firms – especially in the USA, and in a number of smaller European countries outside the EEC – Switzerland, Finland, Norway – all increasing their share within the EEC. The firms based in countries with the largest technological activities – the USA, Japan and FR Germany – had amongst the lowest proportionate increases in foreign technological activities. Most of any increase in foreign technological activities came as a by-product of takeovers and divestitures, rather than an explicit re-location of technological activities.

Finally, this pattern suggests that, globalisation of markets and (increasingly) production notwithstanding, there remains at least one compelling reason for companies to concentrate a high proportion of their technological activities in one location. The development and commercialisation of major innovations requires the mobilisation of a variety of often tacit (person-embodied) skills, and involves high uncertainties. Both are best handled through intense and frequent personal communications and rapid decision-making – in other words, through geographical concentration (see Porter, 1990). In this context, it is worth noting that the rapid product development times in Japanese firms (Clark *et al.*, 1987) have been achieved from an almost exclusively Japanese base, whilst the strongly globalised R&D activities of the Dutch Philips company are said to have slowed down product development.

This is consistent with the interindustry differences in table 3.8, which shows that firms making products with the highest technology intensities are amongst those with the lowest degrees of internationalisation of their underlying technological activities: those producing aircraft, instruments, motor vehicles, computers and other electrical products are all below the average for the population of firms as a whole. In all these products, links between R&D and design, on the one hand, and production, on the other, are particularly important in the launching of major new products, and benefit from geographical proximity.[12]

By contrast, we see in table 3.8 a high proportion of foreign R&D in industries, where some localised technological activities are required, either to adapt products to differentiated local tastes, or to exploit local natural resources: food, drink and tobacco, building materials, mining and petroleum. All in all, these data suggest a general theorem: *localised 'low-tech' products require globalised R & D; global 'high-tech' products do not.*

Table 3.8 *Geographic location of large firms' US patenting activities, according to product group, 1985–1990*
(Percentage shares)

| Product group | Abroad | Of which | | | |
		USA	Europe	Japan	Other
Drink and Tobacco (18)	30.8	17.5	11.1	0.4	1.8
Food (48)	25.0	14.8	8.5	0.1	1.7
Building materials (28)	20.6	9.1	9.8	0.1	1.6
Other transport (5)	19.7	2.0	6.8	0.0	10.9
Pharmaceuticals (25)	16.7	5.5	8.3	1.1	1.7
Mining and petroleum (47)	15.0	9.7	3.5	0.1	1.6
Chemicals (72)	14.4	8.0	5.1	0.3	1.0
Machinery (68)	13.7	3.5	9.1	0.1	1.1
Metals (57)	12.8	5.4	5.7	0.1	1.6
Electrical (58)	10.2	2.6	6.8	0.3	0.4
Computers (17)	8.9	0.1	6.6	1.1	1.1
Paper and wood (34)	8.1	2.4	4.9	0.1	0.7
Rubber and plastics (10)	6.1	0.9	2.4	0.4	2.4
Textiles, etc. (18)	4.7	1.4	1.8	0.8	0.6
Motor vehicles (43)	4.4	0.9	3.2	0.1	0.2
Instruments (20)	4.4	0.4	2.8	0.5	0.8
Aircraft (19)	2.9	0.3	1.8	0.1	0.7
All firms (587)	11.0	4.1	5.6	0.3	0.9

Note:
The parenthesis contain the number of firms in each product group.
Source: Based on data supplied to SPRU by the US Patent and Trademark Office.

3.2 Uneven development amongst firms reflects uneven development amongst countries

They also suggest that conditions in the home country remain preponderant in the technological performance of large firms. This is confirmed in table 3.9, which presents evidence of uneven technological development since the late 1960s amongst the world's largest firms.[13] It shows, in the same 11 technological fields as in table 3.5, trends in the total US patenting of the world's top 20 firms in the period 1985–90. Two major conclusions emerge from this table.

First, the technological strengths and weaknesses of each region, shown in table 3.5, are in general reflected in the number of nationally based large firms appearing in the top 20 in table 3.9.[14] Thus, as summarised in table 3.10, Japanese firms make up 11 of the top 20 firms in motor vehicles and

Table 3.9 *Shares of US patenting for top 20 firms in 11 technical fields: sorted according to shares in 1985–90*

	Nationality	1969–74	1985–90
Fine chemicals			
1 Bayer	FRG	2.84	3.70
2 Hoechst	FRG	1.61	2.53
3 Merck	USA	2.57	2.44
4 Ciba-Geigy	Switzerland	4.33	2.27
5 Imperial Chemical Industries	UK	1.98	1.98
6 E. I. Du Pont De Nemours	USA	1.48	1.79
7 Warner–Lambert	USA	0.71	1.58
8 Eli Lilly Industries	USA	1.67	1.46
9 Dow Chemical	USA	1.21	1.24
10 BASF	FRG	0.58	1.19
11 Pfizer	USA	1.17	1.09
12 American Cyanamid	USA	2.43	1.04
13 Johnson + Johnson	USA	0.36	1.03
14 Boehringer Mannheim	FRG	1.00	1.01
15 Hoffmann–La Roche	Switzerland	1.59	0.96
16 Smithkline Beckman	USA	1.40	0.96
17 Monsanto	USA	2.87	0.88
18 Squibb	USA	1.03	0.88
19 Takeda	Japan	1.21	0.83
20 Beecham	UK	0.23	0.82
Other chemicals			
1 Bayer	FRG	2.57	2.77
2 Dow Chemical	USA	2.40	2.67
3 Hoechst	FRG	2.37	2.45
4 BASF	FRG	1.40	2.31
5 Ciba-Geigy	Switzerland	2.54	1.92
6 General Electric	USA	1.39	1.86
7 E. I. Du Pont De Nemours	USA	3.29	1.68
8 Imperial Chemical Industries	UK	2.14	1.16
9 Shell Oil	Netherlands	0.85	1.09
10 Eastman Kodak	USA	1.33	1.03
11 Union Carbide	USA	1.11	0.86
12 Exxon	USA	0.79	0.84
13 Allied-Signal	USA	1.62	0.81
14 Henkel	FRG	0.35	0.80
15 Rhone-Poulenc	France	0.63	0.69
16 Phillips Petroleum	USA	1.46	0.66
17 Sumitomo Chemical	Japan	0.55	0.65
18 Texaco	USA	0.41	0.64

Table 3.9 (*cont.*)

	Nationality	1969–74	1985–90
19 3M	USA	0.52	0.62
20 Monsanto	USA	2.22	0.61
Materials			
1 3M	USA	1.36	2.26
2 Fuji Photo Film	Japan	0.31	2.24
3 Ppg Industries	USA	2.72	1.81
4 General Electric	USA	2.32	1.76
5 E. I. Du Pont De Nemours	USA	3.48	1.57
6 Hitachi	Japan	0.20	1.43
7 Corning Glass Works	USA	2.77	1.13
8 Dow Chemical	USA	1.26	1.11
9 Hoechst	FRG	1.01	1.08
10 Saint-Gobain Industries	France	0.96	0.95
11 Emhart	USA	0.46	0.93
12 TDK	USA	0.10	0.91
13 Owens-Corning Fiberglas	USA	1.83	0.89
14 Allied-Signal	USA	0.65	0.86
15 Toshiba	Japan	0.38	0.80
16 Sumitomo Electric Industries	Japan	0.06	0.75
17 Kimberly-Clark	USA	0.61	0.75
18 W. R. Grace	USA	0.69	0.71
19 Bayer	FRG	0.59	0.69
20 GTE	USA	0.28	0.65
Non-electrical machinery			
1 General Motors	USA	1.23	0.91
2 Hitachi	Japan	0.18	0.88
3 General Electric	USA	1.22	0.80
4 Canon	Japan	0.05	0.71
5 Toshiba	Japan	0.08	0.69
6 Siemens	FRG	0.29	0.65
7 Philips	Netherlands	0.39	0.54
8 United Technologies	USA	0.47	0.54
9 Nissan Motor	Japan	0.12	0.52
10 Westinghouse Electric	USA	0.56	0.51
11 Honda	Japan	0.03	0.51
12 Allied-Signal	USA	0.78	0.50
13 Toyota Jidosha Kogyo	Japan	0.10	0.50
14 Fuji Photo Film	Japan	0.09	0.41
15 Mitsubishi Denki	Japan	0.03	0.38
16 IBM	USA	0.46	0.38

Table 3.9 (*cont.*)

	Nationality	1969–74	1985–90
17 ITT Industries	USA	0.35	0.35
18 Robert Bosch	FRG	0.24	0.34
19 ATT	USA	0.56	0.34
20 Aisin Seiki	Japan	0.10	0.31
Vehicles			
1 Honda	Japan	0.91	9.12
2 Nissan Motor	Japan	1.74	5.73
3 Toyota Jidosha Kogyo	Japan	0.76	4.84
4 Robert Bosch	FRG	3.79	4.27
5 Mazda Motor	Japan	0.78	2.93
6 General Motors	USA	5.21	2.79
7 Mitsubishi Denki	Japan	0.21	2.78
8 Nippondenso	Japan	1.27	2.57
9 Fuji Heavy Industries	Japan	0.06	2.20
10 Hitachi	Japan	0.38	1.91
11 Yamaha Motor	Japan	0.32	1.85
12 Daimler-Benz	FRG	2.71	1.50
13 Ford Motor	USA	2.41	1.50
14 Brunswick	USA	0.42	1.10
15 Aisin Seiki	Japan	0.34	1.10
16 Lucas	UK	0.95	0.87
17 Porsche	FRG	0.42	0.86
18 Outboard Marine	USA	0.70	0.84
19 Caterpillar	USA	1.99	0.76
20 Kawasaki Jukogyo	Japan	0.08	0.76
Electrical machinery			
1 General Electric	USA	5.77	2.99
2 Westinghouse Electric	USA	3.22	2.68
3 Philips	Netherlands	1.44	2.15
4 Amp	USA	1.10	2.02
5 Mitsubishi Denki	Japan	0.20	1.97
6 Hitachi.	Japan	0.53	1.91
7 Siemens	FRG	1.47	1.85
8 Toshiba	Japan	0.41	1.53
9 General Motors	USA	1.81	1.30
10 GTE	USA	1.20	1.29
11 Motorola	USA	0.51	0.97
12 Matsushita Electric Industrial	Japan	0.85	0.88
13 Asea Brown Boveri Ab	Switzerland	0.83	0.80
14 United Technologies	USA	0.93	0.67

Table 3.9 (*cont.*)

	Nationality	1969–74	1985–90
15 NEC	Japan	0.22	0.63
16 ATT	USA	1.25	0.53
17 Robert Bosch	FRG	0.49	0.53
18 Allied-Signal	USA	0.72	0.52
19 Honeywell	USA	0.76	0.51
20 Canon	Japan	0.07	0.51
Electronic capital goods and components			
1 Toshiba	Japan	0.53	5.29
2 IBM	USA	8.83	5.25
3 Hitachi	Japan	1.71	4.79
4 Motorola	USA	2.15	2.88
5 Texas Instruments	USA	1.97	2.88
6 NEC	Japan	0.97	2.75
7 Mitsubishi Denki	Japan	0.13	2.73
8 Fujitsu	Japan	0.38	2.59
9 General Electric	USA	6.77	2.50
10 Philips	Netherlands	2.81	2.43
11 ATT	USA	6.05	2.09
12 Siemens	FRG	1.75	1.79
13 Honeywell	USA	2.23	1.08
14 Unisys	USA	3.47	1.03
15 Sharp	Japan	0.05	1.03
16 Canon	Japan	0.04	0.97
17 General Motors	USA	1.36	0.92
18 Tektronix	USA	0.26	0.89
19 Thomson-Csf	France	0.32	0.81
20 Sony	Japan	0.43	0.79
Telecommunications			
1 ATT	USA	5.97	4.24
2 Siemens	FRG	2.22	3.25
3 General Electric	USA	4.29	2.89
4 Philips	Netherlands	1.54	2.55
5 Motorola	USA	1.02	2.55
6 NEC	Japan	0.61	2.43
7 Westinghouse Electric	USA	3.30	1.79
8 Toshiba	Japan	0.26	1.64
9 Mitsubishi Denki	Japan	0.21	1.49
10 ITT Industries	USA	3.55	1.41
11 Hitachi	Japan	0.43	1.41
12 General Motors	USA	1.57	1.30

Table 3.9 (*cont.*)

	Nationality	1969–74	1985–90
13 Thomson-Csf	France	0.82	1.26
14 GTE	USA	1.49	1.21
15 IBM	USA	1.19	1.12
16 Northern Telecom	Canada	0.54	1.02
17 Fujitsu	Japan	0.20	0.86
18 Rockwell International	USA	0.63	0.77
19 CGE	France	0.58	0.75
20 Alps Electric	Japan	0.13	0.74
Electronic consumer goods			
1 Canon	Japan	0.95	6.51
2 Fuji Photo Film	Japan	2.12	6.21
3 Eastman Kodak	USA	6.24	3.32
4 Toshiba	Japan	0.42	3.27
5 General Electric	USA	3.97	3.06
6 Philips	Netherlands	2.38	3.04
7 Sony	Japan	1.02	2.94
8 Hitachi	Japan	0.62	2.89
9 Minolta Camera	Japan	0.88	2.47
10 Xerox	USA	3.79	2.29
11 Konica	Japan	0.52	1.95
12 Ricoh	Japan	1.00	1.87
13 Matsushita Electric Industrial	Japan	1.20	1.61
14 Sharp	Japan	0.00	1.60
15 IBM	USA	2.92	1.44
16 Pioneer Electronic	Japan	0.18	1.44
17 Olympus Optical	Japan	0.14	1.22
18 Mitsubishi Denki	Japan	0.07	1.11
19 NEC	Japan	0.36	1.05
20 Siemens	FRG	0.59	1.01
Technologies for extracting and processing raw materials			
1 Mobil Oil	USA	2.17	4.91
2 Exxon	USA	3.00	2.25
3 Halliburton	USA	0.60	1.62
4 Chevron	USA	2.66	1.47
5 Philip Morris	USA	1.32	1.42
6 Baker Hughes	USA	0.41	1.38
7 Texaco	USA	2.49	1.36
8 Phillips Petroleum	USA	2.35	1.36
9 Nabisco Brands	USA	0.32	1.29
10 Amoco	USA	0.18	1.28

Table 3.9 (*cont.*)

	Nationality	1969–74	1985–90
11 Shell Oil	Netherlands	2.13	1.26
12 Allied-Signal	USA	3.00	1.17
13 Atlantic Richfield	USA	0.89	1.10
14 Deere	USA	1.04	0.89
15 Union Oil Of California	USA	0.57	0.83
16 E. I. Du Pont De Nemours	USA	0.87	0.67
17 Nissan Motor	Japan	0.03	0.61
18 Schlumberger	USA	0.85	0.55
19 British-American Tobacco	UK	0.19	0.55
20 Nestle	Switzerland	0.15	0.55
Defence-related technologies			
1 Boeing	USA	1.06	4.29
2 MBB	FRG	1.18	2.50
3 General Electric	USA	1.58	1.44
4 Oerlikon-Buhrle Ag	Switzerland	0.89	1.35
5 British Aerospace	UK	0.66	1.28
6 Morton Thiokol	USA	0.93	1.11
7 Feldmuhle	FRG	1.56	1.08
8 General Dynamics	USA	0.29	1.08
9 Imperial Chemical Industries	UK	1.33	0.95
10 Honeywell	USA	0.31	0.93
11 United Technologies	USA	1.04	0.82
12 Aerospatiale	France	0.35	0.82
13 General Motors	USA	0.54	0.80
14 Westinghouse Electric	USA	0.15	0.80
15 Olin	USA	1.16	0.71
16 Lockheed	USA	0.79	0.69
17 Grumman	USA	0.02	0.66
18 Ford Motor	USA	0.08	0.51
19 Sundstrand	USA	0.02	0.51
20 Rockwell International	USA	0.83	0.46

Source: Based on data supplied to SPRU by the US Patent and Trademark Office.

Table 3.10 *Nationalities of the top 20 firms in US patenting, 1985–1990*

	Japan	United States	West Europe	Correlation of shares of the Top 20: 1969–74 to 1985–90
Defence related technologies	0	14	6	0.37
Fine chemicals	1	12	7	0.54
Industrial chemicals	1	11	8	0.66*
Raw materials based technologies	1	16	3	0.45
Materials	4	13	3	0.41
Electrical machinery	6	10	4	0.68*
Telecommunications	6	10	4	0.70*
Electronic capital goods	8	9	3	0.51
Non-electrical machinery	9	8	3	0.41
Motor vehicles	11	5	4	0.15
Electronic consumer goods	14	4	2	0.27

Note:
* Denotes a correlation coefficient significantly different from zero at 5 per cent level.
Source: Based on data supplied to SPRU by the US Patent and Trademark Office.

14 in consumer electronics and photography, US firms make up 16 of the top 20 in raw materials and 15 in defence, whilst European firms have their largest numbers in chemicals.

Second, in addition to uneven development of large firms in each technological field according to their nationality, there has also been an uneven degree of stability (or instability) in the firms' shares and rankings within each technological field. A casual reading of table 3.9 shows that in some fields, the leaders of the early 1970s continued to be so into the late 1980s, whilst in others new leaders emerged during the period. This is shown statistically in the final column of table 3.10, which presents the correlation of the shares of the top 20 firms in 1985–90 with their shares in 1969–74.

Thus the low (and statistically insignificant) correlations in motor vehicles and in electronic consumer goods, mainly reflect the emergence of Japanese firms as technological leaders in these fields, whilst the high (and statistically significant) correlations in electrical machinery and telecommunications reflect mainly a re-enforcement of the dominance of established US and some European firms. The more stable shares in industrial chemicals reflect the continuing strength of mainly European firms.

4 Possible causes of uneven development

The above.evidence suggests that the behaviour of large firms will not, in and of itself, lead to a wider international spread of R&D and related activities. On the contrary, our statistical analysis confirms Porter's conclusion (1990) that the conditions in large firms' home countries have a major impact on the rate and direction of their technological activities. We shall now propose a framework of analysis that might eventually explain international differences in the rate and direction of technological accumulation. At this stage, it does not lend itself to rigorous modelling and statistical analysis, although it is consistent with the conclusions of a wide range of more qualitative analyses, as well as with our own data.

4.1 International differences in the volume of technological activities: 'institutional failure' in the competence to evaluate and benefit from technological learning

Some of the observed international differences in the volume of technological activities may reflect differences in the degree of market failure.[15] However, we would also stress the importance of *institutional failures* in the competence to evaluate and benefit from investments in technology that are increasingly specialised and professionalised in nature (e.g., industrial R&D laboratories employing highly qualified specialists in a variety of fields of science and engineering), and are long term and complex in their economic impact (e.g., from research on photons, through the laser, to the compact disc, over a period of 25 years). For purposes of exposition, we have found it useful to distinguish between national systems of innovation that we define as *myopic*, and those that we define as *dynamic* (Pavitt and Patel, 1988).

Briefly stated, *myopic* systems treat investments in technological activities just like any conventional investment: they are undertaken in response to a well-defined market demand, and include a strong discount for risk and time. As a consequence, technological activities often do not compare favourably with conventional investments. *Dynamic* national systems of innovation, on the other hand, recognise that technological activities are not the same as any other investment. In addition to tangible outcomes in the form of products, processes and profits, they also entail the accumulation of important but intangible assets, in the form of irreversible processes of technological, organisational and market learning, that enable them to undertake subsequent investments, that they otherwise could not have made.[16]

The archetypal dynamic national systems of innovation are those of FR

Germany[17] and Japan, whilst the myopic systems are the UK and the USA. The essential differences can be found in three sets of institutions:

First, in the *financial* system underlying business activity: in Germany and Japan, these give greater weight to longer-term performance, when the benefits of investments in learning begin to accrue. And they generate both the information and the competence to enable firm-specific intangible assets to be evaluated by the providers of finance (Hu, 1975; Corbett and Mayer, 1991).

Second, there are the methods of *management*, especially those employed in large firms in R&D-intensive sectors: in the UK and USA, the relatively greater power and prestige given to financial (as opposed to technical) competence is more likely to lead to incentive and control mechanisms based on short-term financial performance, and to decentralised divisional structures insensitive to new and longer term technological opportunities that top management is not competent to evaluate (Abernathy and Hayes, 1980; Lawrence, 1980).

Third, there is the system of *education and training*: the German and Japanese systems of widespread yet rigorous general and vocational education provide a strong basis for cumulative learning. The British and US systems of higher education have performed relatively well, but the other two-thirds of the labour force are less well trained and educated than their counterparts in continental Europe and East Asia (Prais, 1993; Newton *et al.*, 1992).

4.2 International differences in fields of technological specialisation: local inducement mechanisms and cumulative trajectories

We further propose that the observed international differences in the sectoral patterns of technological accumulation emerge from the localised nature of technological accumulation, and the consequent importance of the local inducement mechanisms that guide and constrain firms along cumulative technological trajectories. We know from earlier debates about the relative importance of 'technology push' and 'demand pull' that these inducement mechanisms are numerous, and that their relative importance varies amongst sectors. It is nonetheless possible to distinguish three mechanisms.

Factor endowments: examples include the stimulus of scarce labour for labour-saving innovations in the USA; and the different technological trajectories followed by the automobile industries of the USA, and of Europe and East Asia, as consequence of very different fuel prices.

Directions of persistent investment, especially those with strong intersectoral linkages: examples include the extraction and processing of natural resources (N. America, Australia and Scandinavia), defence (USA, France, UK), public infrastructure (France) and automobiles (Japan, Germany, Italy).

The cumulative mastery of core technologies and their underlying knowl-edge bases: examples include Germany in chemicals and machinery, Sweden in machinery, Switzerland in fine chemicals, the Netherlands in electronics; Japan in electronics and automobiles; the USA in chemicals and electronics; the UK in chemicals.

The relative significance of these mechanisms change over time. In the early stages, the directions of technical change in a country or region are strongly influenced by local market inducement mechanisms related to scarce (or abundant) factors of production and local investment opportunities. At higher levels of development, the local accumulation of specific technological skills itself becomes a focussing device for technical change. At this stage, firms become less dependent on the home country for creating the appropriate market signals, and more so for its provision of high-quality skills and knowledge bases that local firms can exploit on world markets.

5 Conclusions

In their essay on the rise and fall of US technological leadership, Nelson and Wright (1992) conclude that two sets of factors – both related to the increasing interdependence of the world economy – led to the erosion of the massive US technological (and productivity) lead held before and immediately after World War Two.

1   Together with massive social changes, increasing international openness of markets after World War Two eroded the US advantages in market size, natural resource availability and more egalitarian income distribution, that were of central importance to the US lead in mass production.
2   The highly educated and increasingly international nature of technological communities has eroded the US lead in 'high-technology' product groups.

If these mechanisms are dominant and can be generalised internationally, we could expect technology gaps to disappear amongst the industrially advanced countries. But both our data and our proposed framework of explanation suggest otherwise.

First (as Nelson and Wright themselves recognise), there exist international disparities in education that will influence (and for a long time to come) the 'human capital endowments' from which firms can benefit. Countries with a strong endowment of science and engineering graduates, but a badly educated workforce are likely to be constrained to specialise in fields like drugs and software, where the skills of the general workforce are not critical. Countries with a skilled general workforce will have a wider range of opportunities, including assembly and process industries, where production-related skills are of central importance.

Second, our evidence suggests that the distinction between myopic and dynamic systems of finance and management does matter, and influences not only the size of the overall commitment of resources to technological accumulation, but the capacity to maintain and develop competencies in core technologies that open a range of potential future product opportunities. Thus, relative overall decline in the USA has been accompanied by a major loss of competence in automobiles and in the UK in automobiles and electronics.

The UK decline is of long standing, and its nature and causes have been widely documented and debated elsewhere. For the USA, Chandler (1992) identifies a number of factors changing US corporate behaviour towards technology and innovation since the 1960s. In particular, the growing influence of business school graduates with universal recipes for management problems, the difficulties facing corporate management in making informed entrepreneurial judgements over a large number of often disparate product divisions, and the changing role of the investment banks from underwriting long-term corporate investments to trading in corporate control, all conspired to change the bases of strategic decisions:

ROI data were no longer the basis for discussion between corporate and operational management as to performance, profit and long-term plans. Instead, ROI became a reality in itself – a target sent down from the corporate office for division managers to meet . . . ROI too often failed to incorporate complex, non-quantifiable data as to the nature of specific product markets, changing production technology, competitor's activity and internal organisational problems – data that corporate and operating managers had in the past discussed in their long person-to-person evaluation of past and current performance and the allocation of resources. Top management decisions were becoming based on numbers, not knowledge. (Chandler, 1992, pp.277–8)

The same features exist in other OECD countries, and there is no reason to believe that they will diminish in future. In our view, uneven and divergent technological development amongst the industrially advanced countries is here to stay.

Notes

This chapter draws heavily on the results of research undertaken in the ESRC (Economic and Social Research Council)-funded Centre for Science, Technology, Energy and the Environment Policy (STEEP) at the Science Policy Research Unit (SPRU), University of Sussex.

1 R&D is a better measure of rates of change in real resources over time, but it measures technological activities in small firms only very imperfectly. US patenting is a better measure of technological activities in small firms and can be broken down quite finely by specific firms and specific technical fields. Neither measure is satisfactory for software technology, but no alternative yet exists. And neither measure captures all the activities that lead to product and process innovations, such as design, management, production engineering, marketing and learning by doing.

2 Government funded R&D performed in industry is excluded. This is concentrated in defense-related activities, and in few countries: principally the US, UK, France and FR Germany, where it has clearly stimulated accumulation in defense-related technologies (see tables 3.5, 3.9 and 3.10 below). Its wider effects on technological accumulation are a matter of debate. Our own conclusion is that defense R&D has considerable opportunity costs, particularly in electronics, where leading-edge technologies and markets have shifted to civilian applications.

3 The correlation coefficient for the 17 countries is 0.82, which is significant at the 5 per cent level.

4 The correlation coefficient for the 17 countries is 0.80, which is significant at the 5 per cent level.

5 If the level and rate of increase of US patenting is a reliable guide to technology levels, S. Korea and Taiwan are now at the level of technology in Japan about 35–40 years ago.

6 See, most recently, Prais (1993).

7 RTA is defined as a country's or region's (or firm's) share of all US patenting in a technological field, divided by its share of all US patenting in all fields. An RTA of more than one therefore shows a country's or region's relative strength in a technology, and less than one its relative weakness. These measures correspond broadly to the measures of comparative advantage used in trade analyses.

8 For this analysis we use a more detailed breakdown than that used in table 3.5. Again on the basis of the US Patent Classification we have divided technologies into 34 fields.

9 Archibugi and Pianta (1992) also show that OECD countries' degree of technological specialisation is increasing over time.

10 National differences in basic research also exist, but appear – in the long term – to adjust to the level of demand for skills and knowledge from the business sector. For further discussion, see Patel and Pavitt, 1994.

11 We use data on the address of inventors patenting in the USA as a proxy measure of the international location of large firms' technological activities. These data are consistent with the available (but less comprehensive) studies based on corporate R & D expenditures. See Patel, 1995.

12 The one technology intensive exception is pharmaceutical products, where the share of foreign R & D is high, but where R & D and production are unimportant compared to the links with high quality basic research, and with nationally based agencies for testing and validation.

•

13 It is based on data that we have compiled on more than 600 of the world's largest, technologically active firms, as measured by their patent activity in the USA (see Patel and Pavitt, 1991a).

14 For a more systematic statistical proof, see Patel and Pavitt, 1991a.

15 In particular, the effects of labour mobility on the incentive for business firms to train their workforce; and the effects of intellectual property rights on the incentive to invest in innovative activities.

16 In other words, investments in technology nearly always have an 'option value'. See Myers (1984) and Mitchell and Hamilton (1988).

17 Sweden and Switzerland have many 'dynamic' institutional characteristics similar to Germany.

4 Technology and growth in OECD countries, 1970–1990

MARIO PIANTA

1 Introduction

Technology has recently become a favourite explanatory factor in growth models. Neo-classical economics has moved from a view of technology as an exogenous variable to the 'new growth theory' which tries to endogenise technology in the model, while maintaining the general equilibrium framework and most traditional assumptions on factors and functions of production and on the behaviour of economic agents, firms and markets. However, the major empirical studies in the neo-classical tradition have adopted a variety of analytical approaches .[1]

These efforts however disregard the fundamental disequilibrating role of technical change which is at the root of the Schumpeterian and evolutionary tradition.[2] This second stream of research has provided a more appropriate framework for investigating growth patterns; it starts from the study of how technology is produced and used and investigates the dynamics of growth with alternative technology-gap models (see Fagerberg, 1987, 1988b, 1994; Dosi, Pavitt and Soete, 1990).

At the same time, a large body of work has recently explored the empirical patterns of technological activities at the country level, investigating international, intersectoral and inter-firm differences. This has led to the definition of 'national systems of innovations' (Lundvall, ed., 1992, Nelson, ed., 1993) as a key concept explaining national specificities in technology (and growth) as a result of particular institutional contexts, firms' organisation, patterns of innovative activity and role of the public sector.[3]

A large gulf exists, however, between the attempt to generalise the relationships between technology and growth in technology-gap models and the investigation of the specificities of national systems of innovation. On the one hand technology-gap models have explored the general impact that technology indicators and other variables – including costs, productivity, investment and organisational factors – have on growth and other measures

83

of performance. These analyses have accounted for the complexity of the real world in ways unmatched by neo-classical models, and have convincingly shown the broad regularities in the link between technology and growth over time, across countries and across sectors (see Fagerberg, 1988b, 1994; Dosi, Pavitt and Soete, 1990; Dalum, 1992, Dosi *et al.*, 1992).

On the other hand, empirical studies suggest that such relationships hold only within particular boundaries of time, country sets and groups of sectors. Exceptions, outliers and special cases are frequent, and the broad regularities in the relationship between technology and growth leave room for persistent specificities. Moreover, as the conditions of observation change (more recent periods are examined, or a broader group of countries is considered), we might expect a change in the relevance and direction of some relationships (and in the models aiming at representing them).

Changed conditions can also be the (endogenous) result of the long-term operation of the relationship between technology and growth. Three decades of generalised catching up by Europe and Japan with the leadership of the US have led a limited set of countries to converging performances in economic and, to a lesser extent, technological activity; both have also become more internationalised and specialised at the sectoral level.[4] In these new conditions we can expect the relationships between technology and growth to change, becoming more complex and different across countries (and sectors).

In fact, the dynamics of innovation and its economic impact are not independent of the particular context which is examined. Technology and growth show different relationships in periods of sustained growth and in recessions, in the 'golden age' of Fordist mass production and in situations of widespread restructuring and de-industrialisation, in a context dominated by engineering and oil-related technologies and in the current diffusion of electronics-based technological systems (Freeman, Clark and Soete, 1982).

Additional 'real world' complications to technology-based models of growth include structural factors such as differences between small open economies and large, protected ones; differences between national institutions with a market-oriented approach, or a state-interventionist one; differences in the nature of technologies, for instance between military and civilian R&D and innovation. Such a variety of contexts makes the efforts at generalising relations difficult, and suggests caution in the model-building exercise.

The relevance of technology for growth – especially 'high-tech' activities – however, is increasingly emphasised in industrial countries. Technology is seen as a source of absolute advantage and 'structural competitiveness' for a nation, as an opportunity for changing the pattern of a country's comparative advantages, and, lastly, as a way of replacing particular inputs used in production. The promises of technological change include higher national shares in world production, lower balance of payments constraints, higher levels of economic activity, a shift towards sectors and activities with higher value added, and greater allocative efficiency (see Dosi, Pavitt and Soete, 1990; Tyson, 1992).

This chapter, empirical in scope, explores the ground between the generalised relationship between technology and growth on the one hand, and the national specificities which emerge in the analyses of national systems of innovation. The chapter firstly investigates regularities and differences found across twenty OECD countries from 1970 to 1990 and, secondly, compares the patterns shown by the three major economies, the United States, Japan and Germany.

2 The cross country study of 20 OECD countries

An appropriate setting for the first part of the empirical analysis is the comparison of twenty OECD countries,[5] a group of economies of different size, which in 1970 had strong differences in their level of development and that by 1990 showed more similar performances; even though significant gaps persist in some less-advanced European countries.

The relationships between technology and growth are complex and multifaceted. A 'virtuous circle' of cumulative causation can be suggested (Lundvall, ed., 1992), where research and innovation sustain a country's technological capability, contribute to its capital accumulation (together with other factors affecting investment), leading to economic growth. In turn, growth provides the resources and incentives for further advances in technology.

In an empirical investigation, technology and growth have to be described using imperfect indicators. Following a large body of studies, as a proxy for the disembodied part of innovative activity we can use data on R&D expenditure (an input indicator) and international patenting (an output indicator). The role of capital accumulation and of the innovations embodied in new plants and machinery can be proxied by data on gross investment. GDP per capita can account for countries' growth performances.

While we can expect a general positive relationship among all these variables, broadly associated to the development process, the link between innovation and investment is less certain. Furthermore, the empirical associations found depend also on the specific form of the variables used. The 'virtuous circle' can operate more or less effectively in different periods and countries; it may be hampered by several factors, such as a mismatch among the aggregate variables, an inappropriate sectoral composition, changed external and institutional conditions – aspects which can only partially be translated into quantitative variables.

No attempt therefore is made here to identify causation or to build a general model of the link between technology and growth. Rather, the aim is to test the stability and limits of simple relationships, their consistency and variations across countries and over time.

> *The variables*: The variables used for the analysis have been identified after a preliminary screening of several technology and economic indicators. The four variables selected provide an accurate description and summarise the patterns common to most other indicators. The variables considered are the following:
>
> *Total R&D expenditure per labour force*, measuring the intensity of the resources used as inputs for the formalised part of innovative activities;[6]
>
> *Patents extended abroad per unit of export*, measuring the inventive output in world markets, taking into account the different size and international orientation of countries;[7]
>
> *Gross fixed capital formation per labour force*, measuring the investment intensity, which is also an indirect indicator of the introduction of innovations embodied in capital goods.[8]
>
> *GDP per capita* (calculated, as the other variables, using purchasing power parities), measuring the level of income, and a proxy for productivity.

Data for the twenty OECD countries have been collected for four five-year periods, 1971–5, 1976–80, 1981–5, 1986–90.[9] In the distribution of these variables the twenty OECD countries show a lower dispersion in investment intensities and GDP per capita than in R&D and (even more) patenting activity. Over the four periods considered, the coefficients of variation show a significant convergence in the technology variables (where countries' differences however remain high), a moderate convergence in GDP per capita in the 1970s and a stable distribution in the 1980s, a moderately increasing divergence since the mid 1970s in the (relatively similar) levels of investment per employee.[10]

Against this background of similarities and differences in the patterns across OECD countries, the associations between economic and technological variables have been investigated. Simple linear correlation coefficients have been calculated across countries for each period, in order to test the stability and the evolution of the relationships over time. Table 4.1 shows the results.

Table 4.1 *Cross-country correlations between innovation and economic variables, 20 OECD countries*

Variables	1971–5	1976–80	1981–5	1986–90
R&D per employee				
Patents per export	0.77	0.79	0.81	0.83
R&D per employee				
Investment per employee	0.57	0.46*	0.52*	0.66
R&D per employee				
GDP per capita	0.86	0.82	0.84	0.84
Patents per export				
GDP per capita	0.77	0.70	0.73	0.72
Investment per employee				
GDP per capita	0.72	0.67	0.72	0.79

Notes:
All coefficients are significant at the 1 per cent level. *Coefficients significant at the 5 per cent level.
Source: CNR-ISRDS, calculations on OECD data, *Main Science and Technology Indicators,* December 1992, *Annual National Accounts,* December 1992.

Total R&D expenditure per employee and patents per unit of export show a high association, confirming the close link between different indicators of innovative activity. The increase in the coefficients over the four periods suggests that OECD countries have moved towards a more common pattern of innovation, with more similar combinations of (aggregate) R&D inputs and patenting output.

R&D per employee and gross fixed capital formation per employee show a positive but lower association, which is particularly weak in the 1976–80 and 1981–5 periods. In years of crisis, restructuring and renewed growth, countries appear to have changed their R&D and capital intensities in different directions, with different roles played by innovation and investment. However, by the late 1980s a close relationship had emerged again.[12]

GDP per capita is strongly and consistently associated to the technology variables (more with R&D and less with patents), and increasingly so with investment intensity.

An important regularity in the relationship between technology and growth across advanced countries is confirmed, with some evidence suggesting that the the crisis of 1976–80 led to different country responses (lowering all coefficients), and that a more common pattern had emerged by the late 1980s when also capital intensity becomes closer to innovation and growth patterns. On the link between R&D and investment one may

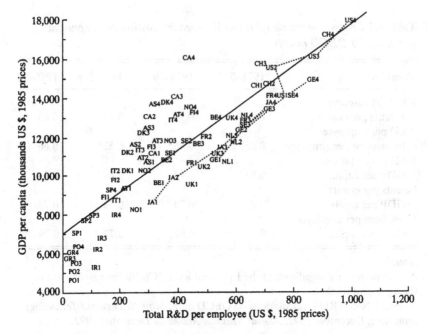

Figure 4.1 R&D per employee and GDP per capita

Notes:

Countries:
AS: Australia	FR: France	NO: Norway
AT: Austria	GE: Germany	PO: Portugal
BE: Belgium	GR: Greece	SE: Sweden
CA: Canada	IR: Ireland	SP: Spain
CH: Switzerland	IT: Italy	UK: United Kingdom
DK: Denmark	JA: Japan	US: United States
FI: Finland	NL: The Netherlands	

Periods: 1: 1971–5
          2: 1976–80
          3: 1981–5
          4: 1986–90

suggest that at first much room existed for national differences, with countries mainly focussing either on the creation of know-how (through R&D) or on the diffusion of innovations (through investment); by the late 1980s however a closer association across OECD countries can be found.

The position of countries: The key relationships of R&D and investment intensities with GDP per capita are highlighted by figures 4.1 and 4.2, which show the distribution along these variables of the 20 countries in the

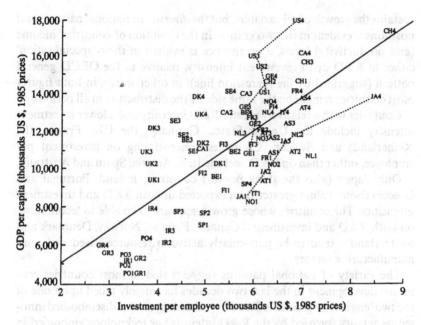

Figure 4.2 Gross Fixed capital formation per employee and GDP per capita
Notes:

Countries: AS: Australia   FR: France   NO: Norway
            AT: Austria   GE: Germany   PO: Portugal
            BE: Belgium   GR: Greece   SE: Sweden
            CA: Canada   IR: Ireland   SP: Spain
            CH: Switzerland IT: Italy   UK: United Kingdom
            DK: Denmark   JA: Japan   US: United States
            FI: Finland   NL: The Netherlands

Periods: 1: 1971–5
           2: 1976–80
           3: 1981–5
           4: 1986–90

four periods, together with the regression lines based on the pooling of all observations. The different paths followed by the three major countries, the US, Japan and Germany are also shown.

The strong link between R&D intensity and GDP per capita is clearly shown in figure 4.1; a slightly more dispersed pattern is found in figure 4.2 for the relationship between investment intensity and GDP per capita;[12] such findings confirm the regularities identified in previous studies (Fagerberg, 1988b; Dosi, Pavitt and Soete, 1990).[13]

What the figures suggest is an underlying process of development which sustains the growth of all variables, but the *diversity* of national patterns and positions is evident in the two graphs. In the evolution of countries' income (and productivity) levels, a persistence is evident in their 'specialisation' either in R&D or in investment intensity, relative to the OECD general pattern (suggested by the regression line); in other words, in both figures most countries remain on the same side of the distribution in all periods.

Countries with a relatively greater R&D intensity and a lower investment intensity include the United States, Germany, the UK, France, the Netherlands and Sweden. Countries concentrating on investment per employee, rather than on R&D, include Italy, Austria, Spain and Australia.

Only Japan (also the three poorest countries, Ireland, Portugal and Greece) shows values greater than expected in both R&D and investment intensities. The countries whose growth appears to be able to 'economise' on both R&D and investment – Canada, Finland, Norway, Denmark and Switzerland – tend to be particularly active in resource-based and non-manufacturing sectors.

The variety of national patterns suggests that in most countries economic development in the last two decades has mainly relied upon one of the two 'engines of growth' offered by technology: *either* disembodied innovative activity (proxied by the R&D intensity) *or* technology embodied in investment (proxied by capital formation per employee).

Such different national paths can now be examined with a time-series analysis focussing on the three major economies, the United States, Japan and Germany, which offer exemplary cases of different patterns. An overview of the evolution of OECD countries is provided in section 4.

3 The time-series analysis of the United States, Japan and Germany

Annual data between 1970 and 1990 for the same variables used so far have been considered for the United States, Japan and Germany and a parallel analysis to that carried out in the previous section across OECD countries is now performed over time for key countries where technology and growth appear to show different relationships.

Table 4.2 shows the correlations coefficients between R&D per employee, external patents per capita (a more appropriate indicator for time-series studies), gross fixed capital formation per employee and GDP per capita.

The close association between technology indicators such as R&D and patenting is confirmed for all three countries by the results of table 4.2.[14] When R&D and investment intensities are compared, however, an important difference emerges, with Japan showing a strong correlation and the US and Germany showing a weaker association.[15]

A high correlation between per-capita GDP and both R&D and patent-

Table 4.2 *Correlations over time between innovation and economic variables, annual data, 1970–1990*

Variables	USA	Japan	Germany
R&D per employee			
Patents per capita	0.83	0.98	0.89
R&D per employee			
Investment per employee	0.67	0.92	0.48*
R&D per employee			
GDP per capita	0.81	0.99	0.99
Patents per capita			
GDP per capita	0.72	0.96	0.85
Investment per employee			
GDP per capita	0.63	0.94	0.47*

Notes:
All coefficients are significant at the 1 per cent level. *Coefficients significant at the 5 per cent level.
Source: CNR-ISRDS, calculations on OECD data, *Main Science and Technology Indicators*, December 1992, *Annual National Accounts*, December 1992.

ing activity is found for all countries (with stronger evidence for Japan). On the other hand, investment per employee is closely associated to GDP per capita in Japan only, with weak correlations for the US and (even lower) for Germany.

In the three major OECD countries changes in GDP per capita are always associated with changes in R&D intensity and in patenting activity, confirming the regularities found in the cross-country study. On the other hand, the investment intensity appears associated to R&D intensity and GDP per capita in Japan only, suggesting that investment in the US and Germany has been more affected by business cycles and has played a different role in the process of growth. Such a diversity may also help explain the weaker associations found among these variables in the cross-country analysis.

The identification of such differences provides a sort of 'measure' of the impact national specificities have on growth. The Japanese pattern of strong, stable and mutually reinforcing relationships between several indicators of technology and growth offers a clear model of 'virtuous circle'. Only Japan shows in figures 4.1 and 4.2 a linear progression in all variables; by 1990 Japan has moved closer to the level of US per-capita income, has reached an R&D intensity and patenting activity similar to the US one, and in terms of investment per employee is considerably ahead of the US.

Germany shows strong economic growth, closely associated to R&D (and patenting), but not to investment intensity. At the end of the two decades Germany has moved closer to the US levels of income, R&D and investment intensity (while showing higher patenting activity).

The United States shows a more erratic pattern and lower associations among all variables, suggesting a much weaker 'virtuous circle' between technology and growth.

The diversity of the national patterns emerging in the past two decades is evident; the different roles played by R&D and investment efforts can now be investigated for the whole group of OECD countries.

4 An overview of OECD countries

How can this evidence on 'virtuous circles' (or the lack of them) between technology and growth be extended to the group of 20 OECD countries? Which country is closer to the US, Japanese or German pattern? Or can other paths of evolution of technological and economic activities be identified?

A simple answer can be found if we summarise countries' growth performances calculating the rates of change of the variables (GDP per capita, R&D and investment intensities) from the first period, 1971–5 to the fourth period, 1986–90. Table 4.3 shows these data as differences from the rate of change calculated over the average of the 20 OECD countries.

OECD countries can be ranked in four groups showing different growth performances:

High growth. Japan and (at a distance) Finland and Ireland are the countries whose strong growth in incomes and productivity is matched by increases in both R&D and investment intensities higher than the average of OECD countries (Norway's growth rate is also affected by the new exploitation of natural resources). In spite of very different national conditions and 'starting points', the countries with the highest gains in GDP per capita show a strong increase in *both* R&D-based innovation *and* investment-embodied technical change, suggesting the operation of a 'virtuous circle'.

Growth above average. Italy, Canada, Austria and Portugal show an above average growth of GDP per capita, sustained by strong increases *either* in R&D intensities *or* (the case of Canada) in investment intensity.

Growth below average. A large group of core European countries, including Germany, the UK, Denmark, Belgium, France and Greece, are close to the average pattern of GDP growth, show a small above

Table 4.3 *Rates of change of innovation and economic variables, 1971–1975 to 1986–1990, in real terms – differences from the rate of change of the average of 20 OECD countries*

Countries	GDP per capita	R&D per employee	Investment per employee
Average of 20 OECD countries	36.2%	59.0%*	12.7%
High growth			
Japan	24.0	76.1	34.7
Norway	26.5	13.3	−6.6
Finland	13.6	119.3	5.1
Ireland	15.8	2.0	1.6
Growth above average			
Italy	11.6	40.4	−3.1
Canada	7.5	−18.6	30.2
Austria	5.3	22.9	−0.1
Portugal	4.3	10.0	−5.3
Growth below average			
Germany	0.2	3.7	−5.3
UK	−0.5	−26.6	12.2
Denmark	−1.6	8.0	−21.1
Belgium	−1.6	−3.3	−1.2
France	−2.9	7.3	−2.2
Greece	−3.7	n.a.	−32.4
Spain	−6.7	146.1	4.5
Sweden	−8.3	52.4	2.5
Relative decline			
Australia	−8.7	−45.9	−2.6
United States	−10.3	−27.7	−1.4
Netherlands	−14.0	−47.7	−25.8
Switzerland	−18.9	−20.0	14.3

Notes:
n.a.: not available. *Average calculated on 19 countries (data are missing for Greece).
Source: CNR-ISRDS, calculations on OECD data, *Main Science and Technology Indicators*, December 1992, *Annual National Accounts*, December 1992.

average increase in one variable and a greater lag behind the OECD pattern in the other.

A peculiar case is that of Spain and Sweden, showing a strong increase in R&D intensity which is however associated to a below average growth in GDP per capita.

Relative decline. The United States, the Netherlands, Switzerland and Australia, with the top income levels, show a decline in all indicators relative to the OECD average. With a mirror image of the high-growth group, countries which have lost ground in the overall pace of technical change (both embodied and disembodied) ended up with a relative decline in their income and productivity levels.

Two patterns should be pointed out here. Firstly, the importance of the catching up process clearly emerges from the evidence of table 4.3. Countries starting with the highest levels not only of GDP per capita, but also of R&D and investment per employee, over the last two decades have often shown a rate of growth lower than the OECD average. This pattern has led to the moderate convergence in most indicators found for the group of twenty countries.

Secondly, in most nations with growth rates close to the average, convergence towards the OECD pattern can be identified. Countries have maintained their relative 'specialisation' in the levels of either R&D or investment intensity, compared with the OECD pattern shown in figures 4.1 and 4.2, but the rate of growth in the variable of national strength has generally been lower than the OECD average increase. Conversely, in the variable where the lag behind the OECD pattern was greater, countries tend to show an above average growth.

This suggests that at first the scope for catching up was so large that countries could 'specialise' either in the production of innovations, reaching high R&D intensities, or in the diffusion and use of technology through higher investment per employee. This was also a result of the strong trade-off countries had to face between the allocation of limited innovative resources between R&D and investment.

As the room for catching up becomes smaller, less diversity in the combinations of R&D and investment intensities emerges, and growth needs to be sustained by a more balanced use of the two 'engines of growth'.

The case of the 'high-growth group', and Japan in particular, whose large gains in income and productivity are associated with strong increases in both R&D and investment intensity, can anticipate a pattern where the marked trade-offs of the past are replaced by increasing complementarities between the different aspects of national technological activities, from research efforts to patenting activity, to diffusion and use of innovations. In fact, in the late 1980s, as shown in table 4.1, many of these indicators have started to show a closer association across OECD countries.

When countries have achieved high-income and productivity levels and are closer to the technological frontier, a more complex set of relationships between technological factors and economic performance is likely to emerge. The 'virtuous circle' of growth can hardly be sustained by concentrating on one particular aspect alone. A new interplay of technological aspects, beyond the R&D and investment indicators used here, is likely to emerge with a greater role played by factors such as knowledge, learning processes, human capital, quality of research, immaterial investment, organisational innovations, favourable institutional conditions, as described by the studies of national systems of innovation[16].

While the analysis is here carried out at the aggregate level, sectoral changes can also play a role in the evolving relationships between technology and growth. In particular, a growing degree of sectoral specialisation has been documented for most countries, and suggests a pattern of increasing national 'division of labour' and competition at the technological frontier[17].

In a context of increasing convergence and competition at the technological frontier, the evidence shows that countries can hardly achieve high growth – and advantages in particular sectors – with efforts focussing either on the production of know-how or on the diffusion of innovations alone. In advanced countries, with the end of the 'catching up era', the regularities in the relationships between technology and growth appear to evolve, while countries show a persisting diversity in national patterns. The 'virtuous circle' of growth is now likely to be the result of a more complex interplay of several knowledge-based aspects of technical change.

Notes

I am indebted to two anonymous referees for their criticism and advice on a previous draft. Patrizia Principessa has assisted in the elaboration of data.

1 On the 'new growth theory' see the reformulation of Grossman and Helpman (1991). Revelant empirical studies on technology and growth include Denison (1967), Griliches (1984), Baumol, Blackman and Wolff (1989), Maddison (1982, 1991). Other studies, such as Scott (1989) and De Long and Summers (1991) have focussed on the role of capital investment in economic growth.

2 Among the vast literature in this approach, see Freeman (1994), Rosenberg (1982), Nelson and Winter (1982), Dosi et al. (1988), A state-of-the-art assessment is in OECD (1992a). Other insights into the long term economic impact of technological systems have come from studies of the long waves in the world economy. See in particular Mensch (1979), Freeman, Clark and Soete (1982), Van Duijn (1983), Freeman, ed. (1984), Reati (1992).

3 The early uses of the concept are in Freeman (1987), Nelson (1987, 1988), Freeman and Lundvall, eds. (1988).

4 On the end of the US technological leadership see, among others, the surveys in Pianta (1988) and Nelson and Wright (1992). On international specialisation see Amendola, Guerrieri and Padoan (1992), Archibugi and Pianta (1997).

5 The countries considered include the United States, Japan, Germany, France, the United Kingdom, Italy, the Netherlands, Belgium, Denmark, Spain, Ireland, Portugal, Greece, Switzerland, Sweden, Austria, Norway, Finland, Australia, Canada.

6 Across the 20 OECD countries, this variable shows a very high association to other key indicators such as total R&D as a share of GDP, industry-financed and performed R&D as a share of GDP, researchers and scientists as a share of the labour force, etc. The R&D intensity per employee has been used in order to make it consistent with the capital intensity indicator.

7 Data on patent applications abroad can be standardized with either population or exports. In a cross country study, it is important to account for the different export orientation of countries, which affects also external patenting patterns, and the latter solution is more appropriate. Using patents per capita would lead to overestimate the activity of small outward-looking economies. In a time series study, on the other hand, there is no need to compare different country structures and exports can show erratic fluctuations; therefore standardizing by population can be more appropriate, and will be used in the next section.

8 Unfortunately data for investment in machinery and equipment only, which would be a better indicator, are not available for all countries and years; when comparisons were possible, the variable used appeared to be a good proxy.

9 Data have been transformed in constant 1985 prices and then converted in US dollars using 1985 purchasing power parities (from OECD, Main Science and Technology Indicators, 1993). The averages for the years included in each period have been calculated. Using five-year averages annual fluctuations are avoided and a more satisfactory picture is provided of interactions which include lagged effects. However, no formal analysis of the lags and feedback loops between technology and growth variables is carried out here.

10 For GDP per capita the coefficients of variation (standard deviation divided by the mean) in the first and last periods are 0.27 and 0.24; for investment per employee values are 0.20 and 0.24; for R&D per employee values are 0.54 and 0.43; for patents per export values are 0.88 and 0.65 (coefficients for the latter two variables are calculated on 18 countries only, due to missing data for Portugal and Greece; estimated values on 20 countries are even higher).

11 In the 1976–80 and 1981–5 periods only a few countries have reduced their R&D per employee, while more than half show levels of investment per employee lower than in the previous period. On the other hand, if we look at R&D and gross fixed capital formation as a share of GDP, a negative relation is found in the 1970s, and no association in the 1980s.

12 In the linear regression estimates, 72 per cent of the variance in GDP per-capita levels is explained by R&D intensities and 52 per cent is explained by gross investment per employee (Data available from the author). When comparing

these results to other studies, it should be stressed that the 20 OECD countries considered here are a more homogeneous group than that considered, for instance, in Fagerberg (1988b).

13 Compared to the clusters identified by Fagerberg (1988b) the evidence provided by Figure 4.1 confirms the presence of a high-R&D, high GDP per capita group (United States, Switzerland, Germany, Sweden) and of the low-R&D, low GDP per capita cluster (Spain, Ireland, Greece, Portugal), while Fagerberg's two intermediate clusters are merged in our distribution.

14 In the three countries the two variables are closely correlated also to the GDP shares of total R&D and industry R&D, and with the share of researchers in the labour force.

15 All countries show clear negative correlations between the *shares* of GDP devoted to R&D and to gross fixed capital formation.

16 On this see, most recently, Abramovitz (1993).

17 The analysis of sectoral specialisation is carried out in Archibugi and Pianta (1992), where a preliminary evidence on the positive impact of greater specialisation on growth is also provided. The parallel patterns of aggregate convergence and sectoral specialisation are examined in Archibugi and Pianta (1997). International competition in advanced technology sectors is examined in Scherer (1992).

5 R&D activity and cross-country growth comparisons

MAURY GITTLEMAN
AND EDWARD N. WOLFF

1 Introduction

Countries whose productive techniques are located away from the world technological frontier can benefit in their attempts to improve efficiency by the fact that such laggards can learn from the leaders. This so-called 'advantage of backwardness', which has been discussed by Gerschenkron (1962) and Kuznets (1973) among others, indicates that diffusion of technology from the leading economies to the more backward ones can potentially be a strong force in enabling productivity levels to converge among countries. Later work, such as Abramovitz (1986), has argued that these potential advantages of backwardness are by no means automatically realised. Other preconditions must be met, among them high levels of investment, the 'social capability' of the country (to use Abramovitz' terminology), the presence of a skilled workforce and appropriate macroeconomic and microeconomic policies.

A further prerequisite, and one whose importance will be assessed in this chapter, may be an adequate level of research and development (R&D) activity. With the requisite scientific and technical effort, nations will find it easier to imitate the technology that they encounter in international trade or through foreign direct investment. As an example, R&D activity might enable domestic firms to realise spillovers from the transfer of technology by multinational corporations.[1] In addition, a country's R&D activity will enhance its ability to make product and process innovations of its own.

Lall (1987) emphasises the former – the adaptation and imitation of technology – rather than the latter – new innovations – in developing what he refers to as technological capability. This is likely to be particularly relevant for developing countries, since it is improbable that movements of the international frontier of technology will occur outside of the industrialised nations. As Lall argues, it requires costly technological efforts for firms to 'change the technological point at which they operate, to gain the knowledge required to assimilate a given technology, to adapt it and to improve on it'. While R&D activity is not the only prerequisite for a country to develop its technological capabilities, clearly investment in R&D and the presence of manpower with sufficient training to engage in R&D are important components of this task.

There is, of course, a voluminous literature that has demonstrated that R&D makes an important contribution to growth at the firm, industry and national level.[2] There is also some recent empirical literature that has examined this issue in a cross-country context. Almost all the cross-national work to date, with two exceptions (at least as far as we are aware) has investigated the relation between productivity growth and R&D among OECD countries only. There are two studies of particular interest. The first, by Coe and Helpman (1993), finds a positive and statistically significant effect of (domestic) R&D stock on total factor productivity (TFP) growth among 22 OECD countries over the period 1971–90. They also include an estimate of foreign (imported) R&D capital stock and find this variable to have a positive and significant effect on TFP growth, particularly among the smaller OECD countries. However, it should be stressed that the authors do not include a catch-up term in their regression. The second, from Working Party No. 1 of the OECD Economic Policy Committee (1993), reports a positive but insignificant effect of the growth in R&D capital stock on labour productivity growth for a sample of 19 OECD countries over four sub-periods between 1960 and 1985 (table 19 of the report). This analysis did include a catch-up term.

Two studies have looked at the effects of R&D on output (GDP) growth. The first, by Fagerberg (1988b), reports a significant and positive effect of the average annual growth in civilian R&D on GDP growth for 22 countries over the 1973–88 period, with a catch-up term included. A second, by Verspagen (1994), uses a cross-country sample consisting of 21 OECD countries, South Korea, and Yugoslavia, and covers the period 1970–85. He finds that with a catch-up term (the initial labour productivity gap) in the specification, the growth in total R&D stock is a positive and significant determinant of GDP growth. Also, distinguishing between business and non-business R&D, he reports that only the former has a statistically significant effect on output growth (the coefficient of non-business R&D is actually negative).[3]

Several studies examine the relation at the industry or sectoral level among OECD countries; two are of particular note. Fecher and Perelman (1989) use data for nine manufacturing industries and 11 OECD countries for the period 1971–86 and find that R&D intensity (the ratio of R&D expenditures to GDP) is a positive and significant determinant of TFP growth in a pooled cross-section regression specification. The estimated

rate of return to R&D is 15 per cent. Dosi, Pavitt and Soete (1990) look at
the relation between labour productivity *levels* and the cumulative number
of patents registered by each country in the US (an index of technological
innovativeness) for 39 industrial sectors. They find that their patent index
is positive and significant for most but not all industrial sectors.[4]

Lichtenberg (1992) is apparently the only study to date which addresses
the role a nation's own R&D activity plays in explaining cross-country
differences in productivity growth among countries at varying levels of
development. He finds a positive and significant coefficient for R&D intensity in explaining growth in GDP per adult between 1960 and 1985. Coe,
Helpman and Hoffmaister (1995) explore the effect of R&D activity in
industrialised countries on TFP levels and growth in a sample of 77 developing countries for the period 1971–90. They find the estimated elasticity
of TFP with respect to foreign R&D capital stocks to be positive and significant and that this marginal benefit from foreign R&D is larger in countries that are more open to foreign trade.

Our chapter considers the relationship between R&D and productivity
growth over the period 1960–88. We rely heavily on the Summers and
Heston (1991) dataset, which covers countries at all levels of development.
We supplement this source with UNESCO data on both R&D intensity
and manpower engaged in R&D.

The present contribution adds to this literature in three ways. First, we
look at the effects of R&D on productivity growth for different groups of
countries, including industrial market economies (OECD countries),
middle-income countries, and lower-income ones. Second, we look at the
effects for different sub-periods – particularly, the 1960s, 1970s and 1980s.
As we discuss below, there are important reasons to believe that R&D
activity may not be as effective in developing countries as in developed
ones. There is also cause to speculate that the effectiveness of R&D varies
over time as the types of technologies on the cutting edge change, along
with the degree to which R&D activity can aid in the imitation and innovation of these technologies. Third, to subject our hypotheses to more
careful scrutiny, we use a measure of manpower engaged in research and
development in addition to the R&D expenditure data that Lichtenberg
employed.

It should be stressed at the outset that we use a standard linear model in
estimating the relationship between productivity growth and R&D. This is
also true for the vast majority of empirical studies on this issue. The main
reason is that even though a more dynamic relationship exists between the
two variables (for example, innovation leads to a growth in demand, which,
in turn, induces more R&D, causing positive feedback effects), data limitations (particularly, the lack of annual time-series data on R&D expendi-

tures for most countries) prevents the estimation of more than a simple specification.

Despite these limitations, we can report two striking results. First, R&D activity is significant in explaining cross-national differences in growth only among the more developed countries. Among the middle income and less-developed ones, the effects are insignificant. Second, and perhaps, more surprising, our analysis suggests that R&D activity has changed in importance over time, in terms of its ability to explain international differences in productivity growth. The exact pattern of this change depends on the sample, but it appears that returns to R&D diminished sharply between the 1960s and 1970s, followed by a modest recovery in the 1980s.

The remainder of this chapter is organized as follows. Section 2 discusses the technology data used and presents descriptive statistics for R&D activity by level of development. Section 3 follows with an econometric analysis of the role of R&D in the convergence and growth of per capita income. The final section offers some concluding remarks.

2 Research and development activity by level of development

We use two variables to measure the degree to which countries employ resources to develop and implement new technologies. The first is expenditure for R&D as a percentage of GNP. The second is scientists and engineers engaged in R&D per capita. Both variables are taken from UNESCO Statistical Yearbooks. The earliest R&D data available are for 1960 and, for the vast majority of countries, such measures are available only for a few scattered years between 1960 and 1988. For the purposes of regression analysis, the two variables were averaged over all available years for the time period of interest. If the measures are stable over time, little bias will be introduced by this procedure.[5]

There are at least two reasons why it is important to consider more than one R&D measure. First, given its intangible nature, R&D activity is inherently difficult to measure. Second, given that countries may differ greatly in terms of the capital intensity of their R&D, the use of a manpower measure may give a different picture of technological activity.[6]

Table 5.1 shows, for both technology measures, averages for four different time periods and for five different samples of countries. The first grouping includes all countries for which the requisite data are available. In addition, we divide the sample into the following four groups, on the basis of World Bank definitions: (1) industrial market economies, (2) upper-middle-income countries, including centrally planned and high-income oil-exporting economies, (3) lower-middle-income countries and (4) low income countries.[7]

Table 5.1 Research and development expenditure and manpower by country group[a] (period averages)

	Ratio of R&D expenditure to GNP (%)				Scientists and engineers engaged in R&D per million of population			
	1960–88	1960–70	1970–9	1979–88	1960–88	1960–70	1970–9	1979–88
All countries								
Mean	0.68	0.73	0.70	0.94	519	503	511	839
Std. dev.	0.66	0.69	0.68	0.87	743	555	718	1064
Coeff. of var.	0.97	0.95	0.96	0.93	1.43	1.10	1.41	1.27
Number	84	58	77	59	85	45	77	64
Industrial market economies								
Mean	1.49	1.30	1.43	1.70	1541	962	1442	1929
Std. dev.	0.66	0.71	0.63	0.69	847	555	777	949
Coeff. of var.	0.44	0.54	0.44	0.41	0.55	0.58	0.54	0.49
Number	19	19	19	19	19	17	19	18
Upper-middle-income, high-income oil exporters and centrally planned countries								
Mean	0.58	0.61	0.54	0.90	474	474	427	812
Std. deviation	0.66	0.73	0.73	0.99	545	452	518	1155
Coeff. of var.	1.15	1.20	1.34	1.09	1.15	0.95	1.21	1.42
Number	20	12	20	15	20	10	19	17

Lower-middle-income countries							
Mean	0.34	0.41	0.35	0.30	124	96	100	205
Std. deviation	0.22	0.35	0.21	0.20	95	57	74	135
Coeff. of var.	0.64	0.86	0.59	0.66	0.77	0.59	0.74	0.66
Number	25	15	21	16	26	11	23	17
Low-income countries							
Mean	0.45	0.34	0.49	0.51	107	70	94	139
Std. deviation	0.36	0.28	0.40	0.46	122	83	85	164
Coeff. of var.	0.80	0.81	0.82	0.92	1.14	1.19	0.91	1.18
Number	20	12	17	9	20	7	16	12

Note:

[a] The division of the countries into groups follows the World Bank convention.

Source: World Bank, *World Development Report, 1986*, table 1, pp. 180–1.

Discerning patterns, both across countries and over time, is made difficult by the fact that the components of the samples do not remain constant over time. This thorny problem – the trade-off between the desire to include as many countries as possible and to maintain a sample that is constant over time – is unfortunately a hindrance in interpreting trends throughout the chapter. Despite this, several are apparent for both R&D measures. For the ratio of R&D expenditures to GNP, it is evident that the richer countries devote a higher proportion of their resources to technological activity – about twice the level of other countries of the world who engage in R&D activity. While this may be expected, since such countries are the centers of technological advance, it may be somewhat surprising that the relationship between per-capita income and R&D expenditure is not that strong. That is, on average, there is little difference among the R&D levels of the other three country groupings, with the low-income countries, in fact, allocating proportionately more of their resources to R&D than the lower-middle-income nations do.

Patterns over time for each of the country groupings, except for the industrialised nations, tend to be obscured by changes in sample size. Among all nations, R&D intensity appears to have remained almost unchanged between the 1960s and 1970s and then shows a marked increase in the 1979–88 period. For the richest nations, there is a pronounced upward trend in R&D intensity between the 1960s and 1980s, while for the other country groups, the patterns are mixed or hard to interpret because of changing sample size.

Another interesting dimension is afforded by the variation in R&D activity among countries, as measured by the coefficient of variation (the ratio of the standard deviation to the mean). The variation in R&D activity is lowest among the industrial market economies and highest among the upper-middle-income countries.[8] Among all countries, the coefficient of variation remains remarkably stable over time, while among low-income, lower-middle-income, and upper-middle-income countries, no clear pattern emerges (partly because of changing country samples). However, among the advanced countries, there is a noticeable downward trend in this dimension between the 1960s and the 1980s, suggesting some convergence in R&D intensity.[9]

The results based on manpower engaged in R&D per million of population are broadly consistent, since the two measures are highly correlated (a correlation coefficient of 0.86 for the two indices averaged over the 1960–88 period). The industrialised nations have by far the highest level of research activity by the manpower measure, followed, in order, by the upper-middle-income countries, the lower-middle-income group, and the low-income group.

All four country groups show an upward trend in R&D manpower intensity between the 1960s and 1980s. For the upper-middle and lower-income countries, R&D manpower per capita appears to have almost doubled. The variation in R&D activity according to the manpower index is again lowest among the industrialised countries and the coefficient of variation shows a clear downward trend for the industrialised countries, while the patterns are again mixed for the other country groups.

3 R&D, the convergence process and growth

In this section, we employ a regression framework to assess the role of R&D activity on the growth in per-capita income among nations. The proper specification for cross-country growth regressions is a matter of some controversy, particularly in light of the findings of Levine and Renelt (1992) that results of such regressions are very sensitive to the specification.

Our dependent variable is the rate of growth of real GDP per capita (RGDP). The main explanatory variable of interest here is R&D intensity. We use the standard specification (see Griliches, 1979, for example) by regressing the growth in RGDP on the ratio of R&D to GNP or, alternatively, on the number of scientists and engineers engaged in R&D per capita.[10] Following previous work (for example, Baumol, Blackman and Wolff, 1989; Barro, 1991; Mankiw, Romer and Weil, 1992; and Wolff and Gittleman, 1993), we include initial per-capita income as a catch-up term, the investment rate and the educational level of the population. The catch-up term has been found to be highly significant in a wide range of specifications, when other variables such as investment or education have been included in the model to allow for conditional convergence.[11] The positive influence of investment as a share of GDP on RGDP growth has been found to be quite robust (Levine and Renelt, 1992). Since our own work (Wolff and Gittleman, 1993) raises questions about the proper measure of the stock of human capital resident in the labor force, we try alternative specifications with a variety of enrollment and attainment rate variables.

An additional justification for including education in our specifications – besides its role as a prerequisite for conditional convergence – is that a more educated workforce may make it easier for countries to adopt and implement new technologies (see Nelson and Phelps, 1966).[12] Since this is a role played by R&D activity as well, it is important to see if measures of R&D remain influential even after controlling for the human capital stock of the workforce.

Our basic estimating equation is then

$$\ln(RGDP_1/RGDP_0)/T = b_0 + b_1\,RGDP_0 + b_2\,INVRATE_{01} + b_3\,EDUCATE + b_4\,R\&D_{01} + \varepsilon \tag{1}$$

where $\ln(RGDP_1/RGDP_0)/T$ measures the annual average growth rate of real GDP per capita from time 0 to 1 (T years), $RGDP_0$ is real GDP per capita at the beginning of the period and $RGDP_1$ at the end of the period, $INVRATE_{01}$ is the average ratio of investment to GDP over the period of analysis, and $R\&D_{01}$ is a measure of average research and development activity during the appropriate time span. As noted above, the catch-up hypothesis predicts that the coefficient b_1 will be negative (that is, countries further behind at the beginning will show more rapid increases in GDP per capita), while b_2 and b_3 should be positive. The coefficient b_4 is also predicted to be positive and its value is usually interpreted as the rate of return to R&D investment.

We employ six alternative measures for the education variable, *EDUCATE*: gross enrollment rates for primary, secondary and higher education and the three corresponding educational attainment rates. Enrollment rates are recorded as of the beginning of the period (or the closest year available), while attainment rates are recorded as of the midpoint of the period under consideration (or the closest year available), in order to more accurately measure the average educational input during the period. $R\&D_{01}$ represents the average level of research and development activity during the period, as measured by R&D expenditures as a proportion of GNP or by scientists and engineers engaged in R&D per capita. Additional details about the sources and methods used in calculating these variables can be found in the footnote to table 5.2.

In the first stage of the analysis, we examine whether research and development is important in explaining cross-country differences in per-capita income growth among countries at different levels of development. The period considered here is 1960–88, since this is the period for which pertinent R&D data are available. While the Summers and Heston (1991) dataset contains data on RGDP for 138 countries for this period, R&D expenditure data are available for only 80 countries and R&D manpower data for 82. Moreover, the inclusion of certain educational attainment variables further reduces the sample size to between 73 and 80 countries (see table 5.2).

Panel A of table 5.2 shows results for R&D intensity, the ratio of R&D expenditure to GNP, from regressions over the 1960–88 period. The specifications differ in terms of the education variable included. Clearly for the total sample of countries, R&D intensity is *not* a significant determinant of per capita income growth. The R&D variable is not significant in any of the cases, and its coefficient is negative in several. The coefficient of the R&D manpower per-capita variable (panel B) is positive in all cases but likewise statistically insignificant in each.

The catch-up effect, as reflected in initial RGDP, is negative and significant at the 1 per cent level in all cases. Investment in physical capital is positive and significant at the 1 per cent level in all equations.

The coefficients of the education variables are all positive. They are significant at the 1 per cent level only for primary enrollment and at the 5 per cent level for secondary attainment when R&D intensity is used. They are not significant in the other cases. This finding raises the possibility that at least some portion of the impact of education on growth, which has been reported in many previous studies, may be attributable to the effects of a more specialised form of human capital, that engaged in R&D, on growth.[13]

It is hard to reconcile our results on R&D expenditures with those of Lichtenberg (1992), who found a positive and significant coefficient for R&D intensity in explaining growth in GDP per adult between 1960 and 1985. Like us, he used the nominal ratio of R&D expenditures to GNP averaged over all years of available data between 1964 to 1989. His sample size was somewhat smaller than ours – 74 countries (though it was smaller for some of his specifications). His model, derived from Mankiw, Romer and Weil (1992) and based on a Cobb–Douglas production function with physical, human and R&D capital, is different than ours, as is his use of non-linear estimation techniques. Despite the differences in data and specification, it is still surprising that his results are so much at odds with ours.

The role of R&D by level of development and period. The impact of research and development on growth is likely to vary by level of development for several reasons. First, R&D tends to be concentrated in manufacturing industries. As a result, those countries in the early stages of industrialisation are likely to spend less on R&D than those with a well-developed manufacturing sector and to find such expenditures less effective as well.[14] Second, a significant portion of the return to a country's R&D expenditures is likely to be in the form of productivity improvements spilling over from the firm actually undertaking the R&D to other enterprises. The information networks needed for firms to take advantage of this public goods aspect of R&D are likely to be more developed in the higher-income nations.

Third, technical progress is more important in enhancing labour productivity for more-developed countries than for less-developed ones. For example, Chenery, Robinson and Syrquin (1986) find that, on average, increases in total factor productivity account for about a third of growth in value added in developing countries and nearly one-half for the developed economies. Since R&D activity is likely to be more important in improving labour productivity by raising technology levels rather than by increasing capital intensity, the lower rates of technical progress in LDCs also suggest reduced scope for R&D activity.

Table 5.2 Regressions of RGDP growth on initial RGDP, the investment rate, educational enrollment and attainment rates, and R&D intensity, 1960–88[a]

Constant	Initial RGDP	INVRATE	Educ. var.	R&D var.	R^2	Adj. R^2	Std. err.	Samp size	Educ. var.
(A) 1960–88 period, ratio of R&D expenditures to GNP									
-0.001	-0.010**	0.130**			0.50	0.48	0.015	80	
(0.79)	(1.03)	(4.77)							
-0.002	-0.010**	0.126**		0.120	0.50	0.48	0.015	80	
(0.36)	(6.13)	(5.55)		(0.42)					
-0.014**	-0.011**	0.081**	0.026**	0.044	0.58	0.55	0.013	80	PRIME65
(2.72)	(7.38)	(3.36)	(3.72)	(0.17)					
0.000	-0.010**	0.093**	0.021	-0.179	0.52	0.49	0.014	80	SECE65
(0.13)	(6.48)	(3.24)	(1.84)	(0.55)					
-0.000	-0.010**	0.114**	0.048	-0.171	0.51	0.49	0.014	80	HIGHE65
(0.11)	(6.22)	(4.78)	(1.45)	(0.49)					
-0.003	-0.010**	0.098**	0.015*	-0.139	0.51	0.49	0.015	73	PRIMA70
(0.64)	(6.22)	(3.48)	(2.10)	(0.42)					
-0.000	-0.010**	0.113**	0.027	-0.250	0.52	0.49	0.015	76	SECA70
(0.20)	(6.29)	(4.49)	(1.72)	(0.67)					
-0.001	-0.010**	0.122**	0.043	0.028	0.50	0.47	0.015	77	HIGHA70
(0.20)	(1.20)	(4.54)	(1.12)	(0.08)					
(B) 1960–88 period, ratio of scientists and engineers engaged in R&D per capita									
-0.000	-0.010**	0.108**		0.424	0.53	0.51	0.014	82	
(0.05)	(6.69)	(4.87)		(1.75)					
-0.012*	-0.011**	0.073**	0.024**	0.292	0.60	0.58	0.013	82	PRIME65
(2.46)	(7.95)	(3.20)	(3.77)	(1.28)					
0.000	-0.011**	0.089**	0.017	0.127	0.54	0.52	0.013	82	SECE65
(0.02)	(6.87)	(3.43)	(1.44)	(0.40)					
-0.000	-0.010**	0.105**	0.026	0.269	0.53	0.51	0.014	82	HIGHE65
(0.09)	(6.63)	(4.60)	(0.81)	(0.87)					

−0.002	−0.010**	0.090**	0.012	0.198	0.54	0.51	0.014	75	PRIMA70
(0.53)	(6.63)	(3.48)	(1.78)	(0.67)					
−0.001	−0.010**	0.110**	0.010	0.277	0.54	0.51	0.014	77	SECA70
(0.21)	(6.55)	(4.57)	(0.57)	(0.73)					
−0.000	−0.010**	0.110**	−0.005	0.473	0.54	0.51	0.014	78	HIGHA70
(0.09)	(6.48)	(4.70)	(0.10)	(1.45)					

Notes:

[a] t-ratios are shown in parentheses below the coefficient estimate.

Dependent variable: $\ln(RGDP_{88}/RGDP_{60})/28$.

Initial RGDP: RGDP per capita at beginning of period, as indicated, measured in units of $10,000s (1985 international dollars).

INVRATE: investment as a share of GDP (both in 1985 international prices) averaged over the appropriate period.

Source for RGDP and INVRATE: Summers and Heston (1991).

PRIME65: Gross enrollment rate in primary school in 1965.

SECE65: Gross enrollment rate in secondary school in 1965.

HIGHE65: Gross enrollment rate in higher education in 1965.

Source for enrollment rates: World Bank (1986, table 29).

PRIMA70: Proportion of the population age 25 and over who have attended primary school or higher in 1970.

SECA70: Proportion of the population age 25 and over who have attended secondary school or higher in 1970.

HIGHA70: Proportion of the population age 25 and over who have attended an institution of higher education in 1970.

* significant at the 5 per cent level, 2-tail test.

** significant at the 1 per cent level, 2-tail test.

Sources and methods for educational attainment: The basic data were obtained from UNESCO, *Statistics of Educational Attainment and Literacy, 1945–74*, Statistical Reports and Studies, no. 22, table 5, 'Educational attainment by age and sex, 1945 onwards', and UNESCO Statistical Yearbook 1990, table 1.4, 'Percentage distribution of population 25 years of age and over, by educational attainment and by sex'. For many countries, no data were available, so missing values were imputed on the basis of a regression of available educational attainment rates for year *t* on educational enrollment ratios (collected from various UNESCO Statistical Yearbooks) for year *t* and for 15 years prior and on per-capita income for year *t*. See Wolff and Gittleman (1993) for details.

R&D/GDP: R&D expenditures as a percentage of GNP, averaged over all available years between 1960 and 1988 (UNESCO Statistical Yearbooks).

SCI/POP: Scientists and engineers engaged in R&D per million of population, averaged over all available years between 1960 and 1988 (UNESCO Statistical Yearbooks).

The issue of changes in the return to R&D over time is somewhat less clear. Perhaps, the major technological development since the 1970s has been the widespread introduction of computers and computer-based equipment. Some (for example, Piore and Sabel, 1984) have gone so far as to argue that this represents a change in technological regime (the 'fourth industrial revolution') from the previous mass production paradigm to one dominated by information technology (IT). We suspect that traditional R&D activity might have less relevance to measured productivity change in an IT-dominated technology, while developments in computer software, factory downsizing and office reorganisation might be more important.

We present two sets of results. In table 5.3, we include all countries with the relevant R&D data in a given time period, while in table 5.4 we include only countries which have the pertinent data for all three time periods. Here our focus is on the two R&D variables, so we show only their coefficients.[15]

Focussing on the 1960–88 period in table 5.3, we find that the two R&D variables are significant only for the industrial market economies. The coefficient of the R&D expenditure variable is quite high for this group of countries – a rate of return of 55 per cent. For the other two groups, the coefficients are insignificant (negative in some cases).[16]

Looking at changes over time for the same group of industrialised countries, we find that R&D intensity is significant only for the 1960–70 period. Moreover, its estimated rate of return is 97 per cent – twice the level estimated for the 1970s or 1980s. The coefficient of the R&D manpower variable is also much greater in the 1960s than in the ensuing two decades. On the basis of the two variables, there is also a slight upward trend between the 1970–9 and 1979–88 periods. The only other case of a significant R&D variable is that for the combined sample of industrial market and upper-middle-income economies in 1960–70, though here the results appear to be dominated by the OECD countries (the total sample size is only thirty).[17] The results for table 5.4 (constant country samples) are very similar.[18]

In sum, our results indicate that the effectiveness of R&D varies by level of development. Only for the most industrialised countries are the two R&D measures significant. The poorer countries do not benefit as much from investments in R&D, presumably because there is less of a need to acquire the latest in technology, the composition of their economies tends to be away from R&D-intensive sectors, and – relative to the richer countries – growth comes more as a result of increases in capital intensity and the reallocation of resources to more productive sectors than from technical progress.[19]

Our regression results also indicate changes over time in the potency of R&D among the industrial market economies. They suggest a decline in effectiveness between the 1960s and the 1970s, with a slight recovery during the 1980s.

4 Concluding remarks

Perhaps our most surprising finding is the fall-off in the return to R&D after 1970 among the advanced economies. We suggested above that this may have been the consequence of the shift in technological regime to IT-based processes. As a result, the avenue to productivity growth is less apt to come from traditional R&D but rather from software-related innovations and the type of industrial restructuring they have spawned.

Some indices of the growth of IT activity in the US are presented in table 5.5. Panel A shows comparative data on the degree of computerisation in the total economy and in manufacturing. Computer investment for year t is measured as the sum of an industry's investment in office, computer and accounting machinery (OCA) over the six years prior to and including t. These investments are in constant (1982) dollars, so that at least a portion of the change in computer quality has been taken into account (see Gordon, 1990). OCA investments were summed over seven years on the grounds that it is a period long enough both to capture most of the computer stock in use and to avoid the undue influence of unusually large or small annual purchases of computers, and short enough to limit the mix of different generations of computers and to avoid the need for making assumptions about depreciation.

Three different measures of computer intensity are used here. The first is the ratio of OCA investments over the six year period to full-time equivalent employment (FTEE), which is shown for four periods: 1954–60, 1964–70, 1974–80, and 1979–85. The first line of table 5.5 documents the tremendous growth in computer investment for the economy as a whole since the 1960s, from $192 per FTEE in 1964–70 to $2,631 per FTEE in 1979–85. This index more than tripled between the 1974–80 and 1979–85 periods.

The second measure is investment in OCA in 1982 dollars per 1982 dollar of GDP and the third is investment in OCA as a proportion of the total capital stock. These results are shown for only the 1974–80 and the 1979–85 periods. Both indices show a tripling in OCA investment intensity between the late 1970s and the early 1980s. Manufacturing was slightly above average in terms of its computer investment but the pattern of growth was almost identical to that of the total economy. These results clearly show the rate of computerisation accelerating substantially during the 1970s and again during the 1980s in US manufacturing and in the economy generally.

Panel B of table 5.5 presents some statistics on the extent to which the US work force performs information functions (though, unfortunately, the series runs only from 1960 to 1980). The basic data come from the 1960,

Table 5.3 Regressions of RGDP growth on initial RGDP, the investment rate, education, and R&D measures for varying country samples over various time periods[a]

	R&D Expenditures/GNP				R&D Manpower per Capita			
	1960–88	1960–70	1970–79	1979–88	1960–88	1960–70	1970–79	1979–88
(A) All Countries								
R&D coeff.	−0.18	0.28	−0.39	0.45	0.13	−0.57	−0.87	0.75*
t-ratio	(0.55)	(0.61)	(0.74)	(1.00)	(0.87)	(0.66)	(1.49)	(2.17)
R^2	0.52	0.38	0.26	0.21	0.53	0.43	0.29	0.26
Adj. R^2	0.49	0.33	0.22	0.15	0.51	0.37	0.25	0.21
Sample size	80	56	77	57	82	44	77	59
(B) Industrial Market Economies								
R&D coeff.	0.55*	0.97*	0.41	0.46	0.55**	0.74	0.03	0.23
t-ratio	(2.31)	(2.27)	(1.02)	(1.55)	(3.11)	(1.46)	(0.09)	(1.19)
R^2	0.68	0.68	0.38	0.30	0.74	0.76	0.33	0.41
Adj. R^2	0.59	0.59	0.20	0.11	0.66	0.68	0.14	0.23
Sample size	19	19	19	19	19	17	19	18
(C) Industrial Market and Upper Middle Income Economies								
R&D coeff.	−0.03	0.89*	−0.25	0.63	0.15	0.20	−0.33	0.69
t-ratio	(0.07)	(2.13)	(0.67)	(1.40)	(0.49)	(0.28)	(0.83)	(1.82)
R^2	0.71	0.62	0.68	0.32	0.77	0.54	0.70	0.40
Adj. R^2	0.67	0.56	0.64	0.24	0.74	0.45	0.66	0.31
Sample size	38	30	39	32	38	26	38	32

(D) Lower Middle and Low Income Economies

R&D coeff.	-0.54	-0.97	-0.06	-1.36	2.89	-6.01	-1.83	3.53
t-ratio	(0.66)	(0.86)	(0.03)	(0.91)	(1.24)	(0.72)	(0.24)	(1.16)
R^2	0.12	0.12	0.08	0.21	0.14	0.11	0.10	0.20
Adj. R^2	0.02	-0.04	0.04	0.05	0.05	-0.16	-0.01	0.06
Sample size	42	26	38	25	44	18	39	27

Notes:

[a] Regressions also include a constant term, initial RGDP, the investment rate, and an education variable. The education variable used for 1960–88 and 1960–70 is the secondary enrollment ratio in 1965. For 1970–79, it is the average of the secondary enrollment ratio in 1965 and that in 1983. For 1979–88, it is the secondary enrolment ratio in 1983. R&D variables are averaged over the corresponding period. The dependent variable is $\ln(RGDP_{t1}/RGDP_{t0})/(t_1 - t_0)$, with years t_1 and t_0 as indicated.

* significant at the 5 per cent level, two-tail test.

** significant at the 1 per cent level, two-tail test.

Table 5.4 Regressions of RGDP growth on initial RGDP, the investment rate, education, and R&D measures for constant country samples over various time periods[a]

	R&D expenditures/GNP				R&D manpower per capita			
	1960–88	1960–70	1970–79	1979–88	1960–88	1960–70	1970–79	1979–88
(A) All countries								
R&D coeff.	0.40	0.35	−0.58	0.97	0.18	−0.53	−0.97	0.81
t-ratio	(1.11)	(0.74)	(0.86)	(1.70)	(0.74)	(0.64)	(0.96)	(1.49)
R^2	0.60	0.54	0.27	0.30	0.44	0.53	0.16	0.19
Adj. R^2	0.55	0.49	0.19	0.23	0.37	0.47	0.05	0.08
Sample size	41	41	41	41	35	35	35	35
(B) Industrial market economies								
R&D coeff.	0.55*	0.97*	0.41	0.46	−0.11	0.35	−0.50	0.35
t-ratio	(2.31)	(2.27)	(1.02)	(1.55)	(0.36)	(0.68)	(0.86)	(1.07)
R^2	0.68	0.68	0.38	0.30	0.53	0.77	0.36	0.29
Adj. R^2	0.59	0.59	0.20	0.11	0.36	0.68	0.12	0.04
Sample size	19	19	19	19	16	16	16	16
(C) Industrial market and upper-middle-income economies								
R&D coeff.	0.69*	0.97*	−0.11	0.94	0.00	0.37	−1.12	0.67
t-ratio	(2.23)	(2.58)	(0.23)	(1.84)	(0.01)	(0.51)	(1.60)	(1.58)
R^2	0.63	0.69	0.42	0.41	0.65	0.57	0.36	0.47
Adj. R^2	0.56	0.64	0.32	0.30	0.57	0.48	0.22	0.35
Sample size	28	28	28	28	23	23	23	23

(D) Lower-middle and low-income economies

R&D coeff.	−1.14	−1.79	0.34	−2.29	16.7	−5.87	22.2	16.3
t-ratio	(0.56)	(1.35)	(0.12)	(0.50)	(2.42)	(0.43)	(1.48)	(1.57)
R^2	0.54	0.49	0.40	0.28	0.58	0.21	0.44	0.27
Adj. R^2	0.30	0.23	0.11	−0.07	0.35	−0.25	0.12	−0.15
Sample size	13	13	13	13	12	12	12	12

Notes:

[a] Regressions also include a constant term, initial RGDP, the investment rate, and an education variable. The education variable used for 1960–88 and 1960–70 is the secondary enrolment ratio in 1965. For 1970–79, it is the average of the secondary enrollment ratio in 1965 and that in 1983. For 1979–88, it is the secondary enrolment ratio in 1983. R&D variables are averaged over the corresponding period. The dependent variable is $\ln(RGDP_{t1}/RGDP_{t0})/(t_1 - t_0)$, with years $t1$ and $t0$ as indicated. A country is included in the sample only if pertinent data are available for all three sub-periods.

* significant at the 5 per cent level, two-tail test

** significant at the 1 per cent level, two-tail test

Table 5.5 *Selected indicators of the growth of information technology (IT) in the United States, 1954–85*

	1954–60	1964–70	1974–80	1979–85
(A) Investment in OCA by Period[a]				
(1) Investment in OCA per full-time equivalent employee (1982 $1,000 per employee)				
a. Total economy	262	192	849	2,631
b. Manufacturing	259	230	1,061	2,948
(2) Investment in OCA per Dollar of GDP				
a. Total economy			0.022	0.063
b. Manufacturing			0.032	0.070
(3) Investment in OCA per Dollar of Capital Stock				
a. Total economy			0.011	0.032
b. Manufacturing			0.017	0.040

	1960	1970	1980
(B) Knowledge and data workers as a percentage of employment by year[b]			
(1) Total economy			
a. Knowledge workers	6.8	7.9	9.1
b. Data workers	35.4	40.6	43.4
c. Information workers	42.2	48.6	52.5
(2) Manufacturing			
a. Knowledge workers	6.2	8.3	8.8
b. Data workers	23.7	26.7	28.3
c. Information workers	29.9	35.0	37.0

Source: [a] Howell and Wolff (1991).
[b] Baumol, Blackman and Wolff (1989), chapter 7.

1970 and 1980 decennial censuses in the US. The tables of raw occupation by industry for each year were first aggregated, in conformity with a relatively consistent classification scheme, into 267 occupations and 64 industries (see Howell and Wolff, 1991, for details). The occupations have been grouped further into three categories: (i) knowledge workers, who are producers of information, (ii) data workers, who are users of information, and (iii) other workers (see Baumol, Blackman and Wolff, 1989, chapter 7, for details). Information workers are defined as the sum of knowledge and data workers.

Professional and technical workers have generally been classified as knowledge or data workers, depending on whether they are producers or users of knowledge. The line is perhaps more than a bit arbitrary at points, and some judgements have been made. Management personnel have been classified as performing a dual data and knowledge function, since their tasks involve both the production of new information for administrative decisions and the use and transmission of this information. Clerical workers are categorised as data workers for obvious reasons. The remaining occupations fall into the 'other worker' category, with some minor exceptions.

Information workers grew from 42 per cent of total employment in 1960 to 49 per cent in 1970 and then again to 53 per cent in 1980. Trends are similar for the two categories of information workers. Knowledge workers increased from 6.8 per cent of the total work force in 1960 to 9.1 per cent in 1980, while data workers rose from 35 to 43 per cent. Information workers as a proportion of the manufacturing work force is, not surprisingly, lower than average. However, this proportion also shows an increase between 1960 and 1980, as well as for knowledge workers and data workers separately. These statistics on labor force composition also indicate a growing importance of IT in the US economy between 1960 and 1980.

There is also evidence from Petit (1993) of a similar increase in IT in a small sampling of OECD countries. He reports, for example, a tremendous growth in the use of micro-electronics between 1976 and 1986 in Belgium, France, Germany, Sweden, the UK and the US. Unfortunately, neither the Petit data nor other sources provide consistent data on IT for enough countries to allow a rigorous econometric test of this hypothesis.

Two other possible explanations of why R&D effectiveness has fallen over time are: (1) a decline in technological opportunities, leading to diminishing returns on R&D investment; and (2) a speed-up in the pace of international spillovers of knowledge, so that while R&D still makes important contributions to productivity growth at a world level, it does not have a measurable impact on relative national growth rates. Each of these possibilities will be explored in turn.

The slowdown in productivity growth in the US during the 1970s spawned a vast literature seeking to determine its causes. Naturally, one important hypothesis warranting examination is that R&D was making less of a contribution to technical progress than previously. The results of Griliches (1980) and others led some to speculate that technological opportunities were being exhausted, reducing the contribution of R&D to growth.[20] More recent work by Griliches and others has since come to an opposite conclusion.[21] As Nadiri (1993) observes, however, a survey of the literature leaves one with the impression that the contribution of R&D to growth may have diminished in the early 1970s and then strengthened in the late 1970s and 1980s. For example, Lichtenberg and Siegel (1991), who analysed the period 1973–85 in the US, find that, while R&D was significant in boosting TFP growth throughout the period, its impact strengthened in the latter half of the period.[22]

Another strand of the technical change literature has taken note of the fact that the ratio of patents to real R&D investment and the number of scientists and engineers (S&E) engaged in research and development has declined in the United States and other developed economies. The patent/S&E ratio in 1990 was at 55 per cent of its level in 1969–70 in the US, 44 per cent in the UK, 42 per cent in Germany, and 40 per cent in France (Evenson, 1993). Whether this implies diminishing returns to R&D is a matter of some dispute. Evenson (1993) argues that there is some evidence that invention potential is being exhausted. Griliches (1989), on the other hand, favors an explanation focussing on a declining tendency to patent.

Thus, while certainly not conclusive, the evidence from more developed countries seems to suggest that changes over time in the returns to R&D form part of the explanation for the patterns on returns to R&D found in our regressions. These findings could reflect the shift in technological regime from one more amenable to R&D-type innovation (mass production) to one less susceptible (IT).

A second potential explanation is that the pace of spillovers of knowledge across borders has quickened, so that R&D is still productive in terms of world growth, but the returns to each country's own R&D has declined. While it is clear that it still takes a major investment for a country to enhance its technological capability, common sense suggests that the increasing integration of the world economy would lead to greater spillovers via multinational corporations and other means. We tried to test this hypothesis directly by including a measure of imported R&D in our regressions. We failed to find support for this hypothesis, but this may be due to limitations in the data.[23] Our result also appears at odds with Coe and Helpman (1993) and Coe, Helpman and Hoffmaister (1995), who did report that foreign R&D capital stocks have a significant effect on domestic total factor productivity growth in both developed and developing countries. Differences in data sources and methodology and specification may account for the disparity in results. Moreover, neither of the Coe *et al.* studies attempts to determine if the pace of international R&D spillovers has quickened over time, nor does any other study that we are aware of. Time-series data collected by the OECD on international payments for patents, licences and technical knowhow would seem to suggest, however, that the pace of international technology transfer has become more rapid.[24]

While anecdotal evidence suggests that these changes over time are due both to temporal changes in the rate of return to R&D investment and to a quickening of the pace of spillover, with the data at hand we cannot distinguish between the two alternatives. Additional research is needed to examine changes over time in the payoff to R&D investment within a broader range of countries and to examine directly the speed at which technology is being transferred across nations, particularly upper-middle-income ones. Data on the spread of IT (or the degree of computerisation) among countries would also allow us to better understand the relationship between the change in technological regime and changes in the measured return to R&D.

Notes

US Bureau of Labor Statistics, Washington, DC; and Department of Economics, New York University, respectively. The authors would like to express appreciation to Daniele Archibugi and two anonymous referees for their comments and to the C.V. Starr Center for Applied Economics at New York University and the Alfred P. Sloan Foundation of New York for financial support. Any opinions expressed here are those of the authors and do not necessarily reflect the views of the US Bureau of Labor Statistics.

1 For a discussion of the transfer of technology by multinational corporations and its connection to productivity convergence, see Blomstrom and Wolff (1994).
2 See, for example, the literature reviews of Griliches (1979) and, more recently, Mohnen (1992).
3 A negative coefficient on government-financed R&D is often reported in single-country studies (see, for example, Wolff and Nadiri, 1993). In the case of the US, this result is often attributed to the high defense component of government R&D. In defense-related work, R&D expenditures usually show up in improved weaponry, which is not reflected in standard productivity measures. In other countries, the negative coefficient may reflect the fact that private industry will finance profitable R&D opportunities out of their own funds, leaving the government to subsidise outlets that may be socially desirable but privately unprofitable.
4 Surprisingly, we could not find any studies containing regressions of productivity growth on R&D by individual industries pooled across countries. Verspagen (1993a), for example, analyses this relation in time-series regressions for individual OECD countries. Interestingly, he finds that the effects of R&D on productivity growth vary by type of industry – most important and significant in high-tech industries and least important and insignificant in low-tech ones.
5 It is well known that, at least in industrialised nations, R&D investment tends to be stable over time, especially in comparison with spending on physical capital. As will be noted below, there may be a tendency for measures of manpower engaged in R&D to rise over time.
6 Lichtenberg (1992), like Verspagen (1994), distinguishes R&D investment by source of funds (private versus government) and also by whether the expendi-

ture is for fundamental or applied research. We do not make such a distinction here because to do so would severely limit our sample size when we conduct our analysis by level of development and time period.

7 The World Bank country groupings change over time, depending on the levels of per–capita income. To standardise the analysis, we use the World Bank's 1986 definitions.

8 The upper-middle-income countries is a diverse group, which includes the newly industrialised countries (NICs), centrally planned economies and high-income oil exporters. However, even when we exclude the latter two types of economies, we continue to find high variation in R&D activity for the group.

9 A similar finding is reported by Archibugi and Pianta (in this volume) for total R&D intensity among 20 OECD countries. However, interestingly, they find no evidence of convergence in R&D intensity on the industry level among OECD countries.

10 Modelling R&D activity at the firm level is rather complicated, as determinants include technological opportunity, appropriability, market structure, firm size and other factors. The situation is even more complex at the national level, given that level of development will be a determinant and that national technology policy and R&D spending will interact. As noted in the Introduction above, modelling national R&D spending is beyond the scope of the chapter and a test of such a model would require data that are not available. Our regression specifications are not meant to suggest, however, that consideration of the forces determining a nation's R&D activity is unimportant. In addition, they are not meant to imply that a country could profitably increase one of the variables without appropriate adjustments of other variables – for example, to speed growth by increasing R&D investment without raising spending on capital equipment that embodies new technology. Rather, our use of this type of specification is aimed at assessing whether – after controlling for other prerequisites for growth such as an educated workforce and high levels of capital investment – R&D acts as a spur to growth.

11 In the parlance of Mankiw, Romer and Weil (1992), 'unconditional convergence' means that less productive countries will grow faster than more productive ones irrespective of any other characteristics of the countries. 'Conditional convergence' implies that given other characteristics (such as the investment rate and the stock of human capital), poorer countries experience faster productivity growth than richer ones.

12 See Benhabib and Spiegel (1992) for a discussion of the Nelson–Phelps model in a cross-country growth context.

13 We also tested to see if the impact of R&D varies by the educational level of the workforce by interacting the secondary school enrollment rate with each of the two R&D measures. To be consistent with later sections, we tried samples with varying country sizes and time periods. In no cases did these interaction variables prove to be statistically significant.

14 An alternative perspective is offered by Pack (1988). Citing evidence from Evenson (1981), he argues that the most important source of technology trans-

fer to LDCs has not been industrial, but rather agricultural technology from international agricultural research centers, which was then adapted by local institutes.

15 The regressions for which the results are shown also contain initial RGDP, the investment rate, and the secondary education enrollment rate as regressors in addition to the R&D variable. The results for the R&D variables change little if other education variables are used or if none is included.

16 We also divided the sample into four quartiles on the basis of per-capita income in 1960 and ran the same set of regressions. The only group for which the R&D variables are significant is the top quartile.

17 Because of the diversity of the upper-middle-income countries, we re-ran the regressions excluding the centrally planned economies and high-income oil exporters. In no case did this change whether or not the coefficient for an R&D variable is significant.

18 The notable exception is that the coefficient on the R&D manpower variable is no longer significant for the industrial market economies over the 1960–88 period. This result is due to the exclusion of Japan, the UK, and the US from the sample.

19 As a referee suggested to us, if the extent of measurement error for the R&D variables is greater at lower levels of development, this could make it more likely that we find R&D to be ineffective for the poorer countries relative to the richer ones. While this possibility cannot be ruled out, the extent of the differences in the coefficients for the developed versus the less-developed samples makes this argument seem unlikely to us.

20 For a more detailed discussion, see Bailey and Chakrabarti (1985).

21 See, for example, Griliches and Lichtenberg (1984), Griliches (1986) and Lichtenberg and Siegel (1991).

22 It should be kept in mind that as Scherer (1983) has argued, continued strong returns to R&D investment do not necessarily mean that technological opportunities are not diminishing, since firms can cut back on R&D expenditures if they do not deem them to be profitable. Unfortunately, we cannot assess this hypothesis in a cross-national context, since a reasonably complete time series on R&D investment is not available for most countries.

23 The variables for imported R&D were calculated as follows: Using UN trade data, for each country we calculated the share of its imports coming from each of its trading partners. These shares, in combination with our R&D measures, were used to compute an import-weighted index of the R&D of the country's trading partners. Given that we did not have measures of R&D for all countries, there is likely to have been substantial measurement error. In addition, to determine the extent to which technology is being imported requires information on the content of the trade.

24 See Nadiri (1993) for further discussion of this point.

6 Aggregate convergence and sectoral specialisation in innovation: evidence for industrial countries

DANIELE ARCHIBUGI AND MARIO PIANTA

Introduction

Technology contributes to the growth of national economies in two apparently contradictory ways. Firstly, technology is a powerful engine of 'disequilibrating' growth for countries able to develop innovation-based competitive advantages. Secondly, technology plays an 'equilibrating' role as the know-how spreads from innovators to imitators allowing laggard countries to catch up. While the former role leads to diverging economic positions of individual countries, the latter contributes to a growing convergence of national performances.

For the relatively small group of advanced industrial countries, economic indicators show that over the past decades a strong convergence has taken place. Among the factors contributing to convergence in income levels, technology plays a prominent role. The first issue addressed in this chapter is whether advanced countries have also converged in their endogenous technological capability. In other words, do indicators of technological intensity show the same pattern of growing similarity among advanced countries as we find for economic variables, or do they reveal strong persisting national differences? And how can patterns of convergence in technology be compared with patterns in economic indicators?

Of course, these very issues assume that technology has some specific properties, as has already been mentioned in chapter 1. In particular, we believe that technology is not a free good easy to use and to be transferred across countries. The ability of nations to benefit from innovations developed elsewhere is certainly conditioned by their general level of infrastructure, including education and capital stock, but also by their efforts formally devoted to scientific and technological activities.

The issue of convergence has mainly been addressed in aggregate terms. However, we believe that it is crucial to consider also the sectoral composition of the technological activities carried out in each country. In fact, the second question here addressed is whether convergence has led to a growing similarity across nations also in their sectors of technological activity. Our hypothesis is that convergence in aggregate innovative levels is combined with a greater technological specialisation and international division of labour.

In order to address such questions, after an overview of the relationship between technology and economic performance in the next section, we examine the patterns shown in the past two decades by aggregate economic and technological indicators (including R&D and patent data). While a broad process of convergence can be found, significant differences among advanced countries do persist in technology, well beyond the reduced distance in terms of GDP per capita and other economic indicators.

In the following section, the sectoral structure of national technological activities, as measured by patents in the US, is considered and the profiles of major OECD countries are compared over two periods of time developing a measure of 'technological distance'. The results show that the partial technological convergence at the aggregate level is rooted in a greater differentiation at the sectoral level, as the degree of national sectoral specialisation has generally increased and countries' innovative efforts have gone in diverging fields. Some conclusions are drawn in the last section, on the link between convergence and specialisation in economic and technological activities.

Technological factors in national economic performances

The issue of convergence and divergence among countries is a major theme in economic research, related to the problems of long-term growth. Many studies have documented the growing economic convergence of a small number of countries, especially in the post-war period.[1] A variety of macroeconomic and international factors – including sustained global demand, expanding industrial capacity, open markets, etc. – have contributed to such an outcome. But a specific role has been played in this process by technology, with the growth of innovative activities and the diffusion of new products and processes incorporating innovations, leading to a rapid dissemination of technical advances and know-how in different fields and countries. These forces can explain the international convergence in productivity (see Dollar and Wolff, 1988). In an empirical analysis which has decomposed the factors contributing to economic convergence, Soete and Verspagen (1993a) have shown that technological change has been the 'driving force for convergence'.

In this way technology has become increasingly important. On the one hand, advanced countries have become more knowledge intensive; on the other hand, the group of countries able to innovate at the frontier of technological opportunities has expanded. International competitiveness, productivity and rates of growth are linked to the ability to innovate successfully, as illustrated by a large number of empirical analyses.[2]

Mastering technological innovation, however, is not a simple task. National success in innovation requires the combination of different factors, including high-quality research, institutions supporting technical advance and adequate managerial skills. Such characteristics define the 'national system of innovation', an important concept for the study of how countries develop and use technology.[3] These factors are unevenly distributed across countries, resulting in substantial differences in the quantity, nature and trajectory of the innovations produced. Since innovation is becoming a key factor in competition among countries, national economic performances are affected by domestic capabilities. In order to keep its position in international competition, each country has to increase its efforts to produce new technology and to adapt and diffuse innovations. Most countries have perceived this changing environment and have devoted a greater attention to science and technology, with a variety of institutional efforts to support technical advance.

One should stress here that patterns in the production of innovations can differ from patterns in the adoption and diffusion of technology. National differences in the efforts and ability to develop innovations on the one hand, and to spread and use them effectively are major factors leading to possible divergence among advanced countries. A key aspect is the ability of national economies to use their different technological resources, and to acquire the know how they do not produce from a variety of channels, including trade, patent cross-licensing, international cooperative research projects and many other specific forms of technology transfer.

Therefore similar economic performances might result from different combinations in the countries' production and use of technology and from different types of specialisation.[4] But the question is to what extent economic convergence can be achieved and sustained without convergence also in innovative activities. We could expect that after some point further progress in economic convergence needs to be sustained by a parallel convergence in countries' ability to carry out research and produce innovations.

In the next section these issues will be explored comparing the patterns of convergence in economic variables, such as per-capita GDP, and in indicators of technological intensity, such as R&D expenditure and patents. While this provides important evidence at the aggregate level, much less attention has been paid to the sectoral composition of innovations in each country. In fact, the convergence in terms of science and technology activities does not inform us about changes occurring in national sectoral strengths and weaknesses. Should we expect that the aggregate convergence is uniform across sectors, or should we expect that each nation is developing its strengths in selected fields? In other words, are countries becoming more similar in the sectoral distribution of their technological activities, or, on the contrary, has the aggregate convergence led to an increasing specialisation at the sectoral level? This issue is of crucial importance for our understanding of international patterns of technological performance, since it shows whether advanced countries are becoming more complementary, leaving more room for international cooperation, or if, on the contrary, they are competing on the same technological fields.

Convergence in economic and technological activities

Four key indicators are used in this section in order to compare patterns of convergence in economic and technological performances over the past twenty years in 20 OECD countries.

Per-capita GDP is the most important and widely used indicator of a country's economic growth, shown in table 6.1. The coefficients of variation (standard deviation divided by the mean) of cross-country data have been calculated for the whole group of 20 countries and for the five most important countries (US, Japan, Germany, United Kingdom and France, which will be referred to as the G5 countries) for the years 1971, 1976, 1981, 1986 and 1990. Data from table 6.1 confirm the widely available evidence on the convergence of income levels among the group of advanced countries: for 20 OECD countries the coefficient of variation falls from 0.29 in 1971 to 0.23 in 1990 (with a small rise in 1986) and the ratio between the richest country (the US) and the poorest (Portugal or Greece) falls from 3.4 to less than 3. The coefficient of variation has also decreased in the G5 countries; this is mainly the result of the rapid catching up of Japan, which ranked fifteenth in 1971 and ranked fifth in 1990. For all countries the greatest increase of convergence in per capita income took place in the 1971–6 period, with the top four countries (US, Switzerland, Canada and Germany) moving very close and remaining so ever since.

The main indicator of a country's technological intensity is the share of GDP devoted to total R&D expenditures. The last decades have seen the spread in most OECD countries of significant R&D activities with the establishment of R&D laboratories as institutions specifically designed to introduce innovations. R&D activities have a twofold aim: on the one hand, to produce innovations, on the other hand, to imitate and adapt knowledge created elsewhere (see Cohen and Levinthal, 1989).

Table 6.2 shows how this variable is distributed in the 20 OECD countries: in 1991 Japan, Switzerland, Sweden, the US and Germany devoted a

Table 6.1 *Gross domestic product per capita (US dollars, 1985 prices)*

Countries	1971	Rank	1976	Rank	1981	Rank	1986	Rank	1990	Rank
United States	13,051	2	14,115	2	15,301	2	16,928	1	18,245	1
Switzerland	14,358	1	14,166	1	15,615	1	16,589	2	17,889	2
Canada	10,145	6	12,316	3	13,778	3	15,195	3	16,249	3
Germany	10,324	5	11,715	4	13,049	4	14,196	4	15,597	4
Japan	7,884	15	9,089	15	10,834	14	12,486	10	15,003	5
France	10,089	7	11,350	6	12,572	5	13,412	7	14,841	6
Sweden	10,406	3	11,677	5	12,181	6	13,562	6	14,363	7
Denmark	9,896	9	11,054	9	11,545	10	13,749	5	14,258	8
Austria	8,607	12	10,296	11	11,584	9	12,536	9	14,143	9
Finland	7,994	14	9,364	14	10,814	15	12,249	12	14,006	10
Belgium	8,923	11	10,610	10	11,510	11	12,240	13	13,962	11
Australia	10,369	4	11,310	7	12,180	7	13,261	8	13,869	12
Italy	8,184	13	9,576	13	11,252	13	12,182	15	13,641	13
The Netherlands	9,927	8	11,219	8	11,666	8	12,297	11	13,401	14
United Kingdom	8,996	10	9,950	12	10,450	16	12,033	16	13,400	15
Norway	7,710	16	9,040	16	11,291	12	12,217	14	12,309	16
Spain	6,379	17	7,728	17	7,752	17	8,383	17	10,035	17
Ireland	4,982	18	5,708	18	6,821	18	7,248	18	9,116	18
Portugal	4,170	20	4,746	20	5,530	20	5,888	20	7,103	19
Greece	4,189	19	5,130	19	5,626	19	5,962	19	6,256	20
Avg. G20	8,825		10,008		11,068		12,131		13,384	
Coeff. Var. G20	0.29		0.25		0.24		0.25		0.23	
Max/min	3.44		2.98		2.82		2.87		2.92	
Avg. G5	10,053		11,244		12,441		13,811		15,417	
Coeff. Var. G5	0.17		0.15		0.14		0.13		0.10	
Max/min	1.66		1.55		1.46		1.41		1.36	

Note:
G5 includes: United States, Japan, Germany, France and United Kingdom.
Source: CNR-ISRDS, elaboration from OECD data.

share close to 3 per cent of GDP to total R&D activities, while the rest of the countries have values spreading from 2.4 per cent (France) to 0.5 per cent (Portugal and Greece). Convergence among the 20 OECD countries has steadily increased also in this indicator, with the coefficient of variation falling from 0.53 in 1971 to 0.43 in 1990; a marked fall is also found in the ratio of the highest to the lowest value, from 25 to 6. The Japanese catching up should again be noted, from the seventh most R&D intensive economy in 1971 it has become the leading country twenty years later. The

Table 6.2 *Total R&D expenditure as a percentage of GDP*

Countries	1971	Rank	1976	Rank	1981	Rank	1986	Rank	1991	Rank
Japan	1.71	7	1.80	6	2.13	6	2.56	5	2.88[n]	1
Switzerland	2.33	2	2.45	1	2.29	5	2.88	3	2.86[m]	2
Sweden	1.49	8	1.79[c]	7	2.30	4	2.89[i]	2	2.85[m]	3
United States	2.47[c]	1	2.30	2	2.45	1	2.91	1	2.82	4
Germany	2.20	3	2.16	4	2.43	2	2.73	4	2.81[n]	5
France	1.88	6	1.75	8	1.97	7	2.23	7	2.40[n]	6
United Kingdom	2.10[c]	4	2.17[e]	3	2.41	3	2.34	6	2.27[m]	7
The Netherlands	2.06	5	1.97	5	1.88	8	2.22	8	2.16[m]	8
Finland	0.87	14	0.91[e]	15	1.19	12	1.67	10	1.88[n]	9
Norway	1.10[b]	12	1.34[e]	9	1.29	10	1.62[i]	11	1.87	10
Belgium	1.41	9	1.33[e]	10	1.62[h]	9	1.68	9	1.69[n]	11
Denmark	0.96[b]	13	0.99	13	1.10	14	1.32	13	1.54[m]	12
Austria	0.61[b]	17	0.92[e]	14	1.17	13	1.31	14	1.50	13
Canada	1.36	10	1.04	11	1.23	11	1.46	12	1.43	14
Italy	0.85	15	0.77	17	0.87	16	1.13	16	1.35	15
Australia	1.22[d]	11	1.00	12	1.00	15	1.27	15	1.23[l]	16
Ireland	0.78	16	0.83[e]	16	0.73	17	0.89	17	0.88[n]	17
Spain	0.27	19	0.34	18	0.42	18	0.61	18	0.87	18
Portugal	0.38	18	0.27	19	0.35[g]	19	0.45	19	0.50[l]	19
Greece	0.10[a]	20	0.18[f]	20	0.21	20	0.33	20	0.47[m]	20
Avg. G20	1.31		1.32		1.45		1.72		1.81	
Coeff. Var. G20	0.53		0.52		0.50		0.47		0.43	
Max/min	24.70		13.61		11.67		8.82		6.13	
Avg. G5	2.07		2.04		2.28		2.55		2.64	
Coeff. Var. G5	0.13		0.11		0.08		0.10		0.09	
Max/min	1.44		1.31		1.24		1.30		1.27	

Note:
G5 includes: United States, Japan, Germany, France and United Kingdom.
Source: CNR-ISRDS, elaboration from OECD data.
[a] estimates; [b] 1970; [c] 1972; [d] 1973; [e] 1975; [f] 1979; [g] 1982; [h] 1983; [i] 1985; [l] 1988; [m] 1989; [n] 1990.

US, on the contrary, fell from the first to the fourth position. Within the group of G5, the process of convergence has stopped during the 1980s since France and the UK lagged behind Japan, the US and Germany.

Looking at the pattern of convergence of smaller groups of countries, a strong and sustained convergence is found for the top five countries in terms of R&D intensity (Japan, Switzerland, Sweden, the US and Germany), while France, the UK and the Netherlands, in spite of an early

Table 6.3 *Industry financed R&D as a percentage of domestic product of industry*

Countries	1971	Rank	1976	Rank	1981	Rank	1986	Rank	1990	Rank
Japan	1.19	3	1.25	3	1.55	3	2.00	3	2.36	1
Sweden	1.13	5	1.57^f	1	1.92	1	2.60^l	1	2.35^o	2
Germany	1.38	1	1.40^r	2	1.76	2	2.12	2	2.12	3
United States	1.13	4	1.14	6	1.35	4	1.63	4	1.52	4
Finland	0.59	10	0.66^f	10	0.83	9	1.24^l	8	1.51^o	5
Belgium	0.79^d	8	1.19^h	4	1.27	6	1.46	5	1.45	6
United Kingdom	1.05^c	6	1.09^r	7	1.29	5	1.41	6	1.41	7
The Netherlands	1.28	2	1.19	5	1.09	7	1.36	7	1.37^o	8
France	0.85	7	0.92	8	1.06	8	1.21	9	1.30^o	9
Norway	0.54^c	11	0.68^f	9	0.65	13	1.06^l	10	1.09^o	10
Denmark	0.60^b	9	0.59	11	0.72	11	0.97	11	1.08^o	11
Austria	0.36^b	15	0.55^r	12	0.75	10	0.81^l	12	0.83	12
Canada	0.53	13	0.42	14	0.67	12	0.80	13	0.75^m	13
Italy	0.53	12	0.45	13	0.52	14	0.55	14	0.73	14
Ireland	0.40	14	0.29^f	15	0.36	15	0.54	15	0.68	15
Australia	0.25^d	16	0.21^g	16	0.20	17	0.47	16	0.51^o	16
Spain	0.13^a	17	0.21	17	0.21	16	0.36	17	0.42^o	17
Portugal	0.10^a	18	0.05^g	18	0.12^l	18	0.14	18	0.15^n	18
Avg. G18	0.71		0.77		0.91		1.15		1.20	
Coeff. Var. G18	0.55		0.58		0.58		0.55		0.52	
Max/min	13.80		31.40		16.00		18.57		15.73	
Avg. G5	1.12		1.16		1.40		1.67		1.74	
Coeff. Var. G5	0.15		0.14		0.17		0.21		0.24	
Max/min	1.62		1.52		1.66		1.75		1.82	

Note:
G5 includes: United States, Japan, Germany, France and United Kingdom.
Source: CNR-ISRDS, elaboration from OECD data.
^{a} estimates; ^{b} 1970; ^{c} 1972; ^{d} 1973; ^{e} 1975; ^{f} 1977; ^{g} 1978; ^{h} 1979; ^{i} 1982; ^{l} 1985; ^{m} 1987; ^{n} 1988; ^{o} 1989.
Data for Switzerland and Greece not available.

start at the top of the 'convergence club', have diverged since 1981, falling behind the leading group. The remaining countries have generally increased their convergence, but remain at a great distance from the top countries. The aggregate evidence for the 20 OECD countries therefore conceals marked differences both in the level of national technological intensity and in the direction of the changes under way.

In table 6.3 the R&D which is financed by and performed in industry (thus excluding government funded projects, including military R&D) is considered; this portion of R&D expenditure is more directly linked to economic growth than government funded R&D. The ranking of countries changes substantially: Japan, Sweden and Germany (and probably Switzerland, but data are not available) have a share greater than 2 per cent of the domestic product of industry, while the US follows with 1.5 per cent and the remaining countries are widely scattered behind. Among the 18 countries considered here, evidence of convergence is much weaker than for previous variables. The coefficient of variation increases until 1981 and falls slightly only in the last decade; differences among countries remain extremely strong.

Within the group of G5 a marked process of divergence is found. Japan and Germany show a stable pattern of common growth, while the US increases its divergence: although the US ranks fourth both in 1971 and in 1990, its absolute distance from Japan and Germany has widened.[5] The other countries remain at a distance and show a little convergence only in the last period considered.

The differentiated pace of convergence in business funded and total (including government funded) R&D shows that while governments of laggard OECD countries gave a high priority to the development, with public funds, of national scientific and technological capabilities, the business sector in most countries has been reluctant to move towards technology-intensive production.

The fourth indicator considered here is the number of patent applications extended abroad by national inventors divided by a country's exports. Extended patents are an indicator of the production of innovations and are directly linked to their international diffusion. They need to be related to a measure, such as total exports, of the size and the international orientation of a national economy.[6] Table 6.4 presents such data, showing a strong and continuing process of convergence, evenly spread across all 20 OECD countries, with a slight slowdown in the 1980s, but with extremely high differences across countries even in 1990. This indicator also shows a remarkable convergence for the G5 group.

Many other indicators have been examined in order to describe the patterns of national convergence and divergence in technological activities (see Archibugi and Pianta, 1992), but the three described above provide a reasonable picture of the patterns that can be identified in three key areas: aggregate indicators of total national research efforts, the activity of firms in industrial innovation and the evidence from output indicators.

Figures 6.1 and 2 summarise the evidence of convergence in economic and technological indicators for the 20 OECD countries and the G5 group, showing the values of the coefficient of variation from 1971 to 1990, cal-

Table 6.4 *External patents per unit of exports (Patent applications at foreign patent offices per million US$ of exports at 1985 prices)*

Countries	1971	Rank	1976	Rank	1981	Rank	1986	Rank	1990	Rank
Switzerland	1.06	1	0.67	1	0.58	1	0.57	1	0.57	1
Japan	0.40	5	0.27	5	0.30	5	0.37	5	0.44	2
United States	0.83	2	0.44	2	0.45	2	0.53	2	0.44	3
Finland	0.14	12	0.14	9	0.16	9	0.23	8	0.41	4
Sweden	0.42	4	0.32	3	0.40	3	0.40	3	0.38	5
Germany	0.53	3	0.32	4	0.36	4	0.37	4	0.36	6
France	0.31	6	0.20	6	0.20	6	0.24	7	0.26	7
United Kingdom	0.27	7	0.16	7	0.19	7	0.21	9	0.26	8
Australia	0.11	13	0.10	12	0.18	8	0.26	6	0.23	9
Denmark	0.20	9	0.14	10	0.15	10	0.20	10	0.21	10
The Netherlands	0.19	10	0.11	11	0.13	12	0.13	12	0.15	11
Austria	0.22	8	0.15	8	0.15	11	0.18	11	0.15	12
Italy	0.14	11	0.10	13	0.10	13	0.13	13	0.12	13
Norway	0.10	15	0.06	15	0.07	14	0.07	14	0.11	14
Canada	0.10	14	0.08	14	0.06	15	0.06	15	0.06	15
Belgium	0.08	16	0.04	17	0.04	17	0.05	16	0.05	16
Ireland	0.04	18	0.05	16	0.05	16	0.03	17	0.05	17
Spain	0.05	17	0.04	18	0.03	18	0.03	18	0.04	18
Greece	0.00	20	0.01	19	0.01	19	0.01	19	0.02	19
Portugal	0.01	19	0.00	20	0.00	20	0.01	20	0.00	20
Avg. G20	0.26		0.17		0.18		0.20		0.21	
Coeff. Var. G20	1.03		0.94		0.86		0.81		0.76	
Max/min[a]	23.65		19.11		16.97		19.41		13.26	
Avg. G5	0.61		0.50		0.69		0.89		1.17	
Coeff. Var. G5	0.48		0.46		0.48		0.44		0.34	
Max/min	0.30		3.64		3.19		2.71		2.17	

Note:
G5 includes: United States, Japan, Germany, France and United Kingdom.
Source: CNR-ISRDS, elaboration from OECD data.
[a] Excluding Greece and Portugal.

culated for the four variables discussed above. Countries differ in terms of per-capita GDP much less than in terms of any technological indicator. They have converged more in terms of total R&D intensity than in terms of the R&D which is financed and performed by industry. More importantly, a significant divergence has taken place within the G5 group in terms of industrial R&D. Extremely high differences remain across OECD countries in terms of an output indicator such as international patenting per

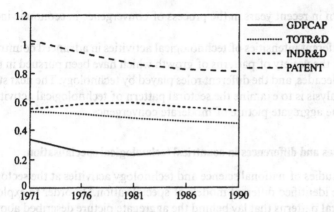

Figure 6.1 Economic and technological indicators, coefficient of variation for 20 OECD countries

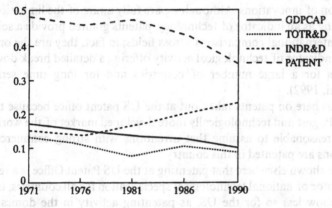

Figure 6.2 Economic and technological indicators, coefficient of variation for five major countries

unit of exports, although, from a dynamic perspective, it has strongly converged for both the G20 and the G5 groups.

Over the past 20 years a small increase of convergence in per-capita GDP can be found, and cross-country differences in industry R&D have not decreased significantly, while convergence is increasing for total R&D and, more rapidly, for patenting performances. While a general 'catching up' in technological activities has taken place in the group of OECD countries in the past two decades, very strong differences do persist in the ability to carry out research and to produce patented inventions. And one may wonder whether such continuing differences in technology have contributed to the slow down in recent years in the process of convergence in economic indicators.

The different intensities of technological activities in advanced countries point out the variety of patterns of growth which have been pursued in the last two decades, and the different roles played by technology. The next step in the analysis is to examine the sectoral pattern of technological activities behind the aggregate picture of moderate convergence.

Similarities and differences in countries' technological specialisation

Several studies of national science and technology activities at the sectoral level have identified different models of specialisation.[7] In order to explore the sectoral patterns that lay behind the aggregate picture described above, we need to ask whether the sectoral composition of national technological activities has become more or less similar across advanced countries.

Patents granted in the US are taken as an indicator of countries' sectoral distribution of innovations. Although we are fully aware of the limitations of patenting as an indicator of technology, patents granted provide a solid base for international comparisons across fields; in fact, they are the only available indicator of technological activity offering a detailed break down by sectors for a large number of countries and for long time series (Archibugi, 1992).

We focus here on patents taken out at the US patent office because the US is the largest and technologically more developed market of the world, and it is reasonable to assume that inventions with strong commercial expectations are patented in this country.

We have shown elsewhere that patenting at the US Patent Office is a reliable indicator of national technological specialisation for all countries, but it is somehow less so for the US, as patenting activity in the domestic market has a sectoral distribution which is slightly different from that emerging in foreign markets (see Archibugi and Pianta, 1992). Particular caution is therefore needed in interpreting the results for the US.

In order to address the question of how similar or different national specialisations are, the sectoral distribution of technological activities has been compared, developing a measure of distance between pairs of countries. The distance among individual countries will be analysed for the periods 1975–81 and 1982–8, using data on the sectoral distribution across 41 SIC classes of patents granted in the US.[8]

Since we intend to compare the composition and not the level of national inventions, countries' patents need to be standardised and their percentage distribution will be considered (countries' shares of total patents granted in the US in 1990 range from 53 per cent for the US, to 0.5 per cent for Spain.

Other OECD countries with a lower number of patents have been excluded since they do not provide statistically reliable information). The method we have used is an application of the chi square statistic. Firstly, the percentage distribution of patents in 41 non-residual SIC classes (see appendix 1 for their listing) is considered for two countries. The square of the difference between these percentages is calculated for each class, and divided by the share the class holds in the world total patents; the sum of these weighted squared differences is the distance indicator used in our analysis. The formula used is the following

$$D_{ab} = \left\{ \sum_{i=1}^{n} (p_{ia} - p_{ib})^2 / p_{iw} \right\} / D_{max} \times 1,000$$

where:

D_{ab} is the distance between country a and country b;

p_{ia} is the percentage of patents of country a in sector i;

p_{ib} is the percentage of patents of country b in sector i;

p_{iw} is the percentage of patents of the world total in sector i;

n is equal to 41 non-residual SIC classes.

D_{max} is the maximum value of the distance for a given world distribution in n classes.

The distance index between two countries is equal to zero when they have the same percentage distribution of patents across classes, and it grows rapidly when one country is strong in fields where the other holds few or no patents. Since the propensity to patent (i.e., the number of patents for each unit of innovative activity) varies considerably across sectors (Scherer, 1983), the method employed here compares the number of patents in each class for each country with the number of patents in the same class for another country.

Table 6.5 provides the distance indexes for 13 major OECD countries. The pattern of relative differences emerging from the matrix is fairly complex; the US has the closest similarities to France, the UK and Canada in both periods. Japan is closest to the US and France, but with higher distance indexes. Germany has a sectoral distribution of patents similar to France, the UK and Italy.

In turn, France and the UK are very close together and are similar to the US and Germany. Italy is close to Germany, followed at a distance by France, the UK and Switzerland. The remaining countries have very high distance indexes and tend to show closer similarities either to the US or the major European countries.

In interpreting such patterns of relative differences across countries the impact of countries' size of patenting activities is evident. The larger coun-

Table 6.5 Distance among the patterns of technological specialisation of advanced countries

Countries	Period	USA	Japan	Germany	France	United Kingdom	Italy	The Netherlands	Belgium	Denmark	Spain	Switzerland	Sweden	Canada
USA	1	—	22.33	17.40	8.95	12.25	37.00	28.82	46.71	36.13	68.51	56.40	31.85	14.19
	2	—	24.64	17.11	6.89	11.02	35.18	23.36	45.11	37.20	69.06	46.09	29.75	13.84
Japan	1	22.33	—	29.28	23.70	25.79	50.74	43.18	57.48	56.14	104.10	73.21	68.58	56.14
	2	24.64	—	37.93	25.61	30.18	60.86	35.26	85.40	79.02	104.38	76.13	64.80	79.02
Germany	1	17.40	29.28	—	9.62	9.51	13.36	54.04	31.20	42.27	45.05	27.89	43.58	31.89
	2	17.11	37.93	—	12.24	14.74	13.58	48.81	46.09	132.19	40.03	24.02	29.18	28.87
France	1	8.95	23.70	9.62	—	3.24	23.22	32.78	40.09	39.13	51.03	40.04	39.63	22.76
	2	6.89	25.61	12.24	—	5.09	26.53	25.47	41.04	40.85	55.71	35.78	33.78	21.85
United Kingdom	1	12.25	25.79	9.51	3.24	—	26.44	78.24	37.50	33.77	53.78	39.02	42.40	27.86
	2	11.02	30.18	14.74	5.09	—	24.37	37.15	33.43	35.33	56.52	32.14	38.55	28.65
Italy	1	37.00	50.74	13.36	23.22	26.44	—	65.36	42.30	44.93	40.49	29.08	63.98	45.67
	2	35.18	60.86	13.58	26.53	24.37	—	64.82	53.61	41.20	38.19	19.79	44.79	42.18
The Netherlands	1	28.82	43.18	54.04	32.78	78.24	65.36	—	85.78	75.04	105.82	82.71	79.91	46.56
	2	23.36	35.26	48.81	25.47	37.15	64.82	—	65.38	74.14	119.77	74.82	73.68	41.17
Belgium	1	46.71	57.48	31.20	40.09	37.50	42.30	85.78	—	64.95	84.88	47.47	86.04	68.43
	2	45.11	85.40	46.09	41.04	33.43	53.61	65.38	—	59.65	90.47	53.45	77.72	61.77
Denmark	1	36.13	56.14	42.27	39.13	33.77	44.93	75.04	64.95	—	57.73	53.34	41.58	41.24
	2	37.20	79.02	132.19	40.85	35.33	41.20	74.14	59.65	—	48.95	47.76	36.32	36.09
Spain	1	68.51	104.10	45.05	51.03	53.78	40.49	105.82	84.88	57.73	—	51.99	63.30	55.79
	2	69.06	104.38	40.03	55.71	56.52	38.19	119.77	90.47	48.95	—	60.10	42.30	56.86

Switzerland	1	56.40	73.21	27.89	40.04	39.02	29.08	82.71	47.47	53.34	51.99	—	76.81	79.79
	2	46.09	76.13	24.02	35.78	32.14	19.79	74.82	53.45	47.76	60.10	—	87.65	64.02
Sweden	1	31.85	68.58	43.58	39.63	42.40	63.98	79.91	86.04	41.58	63.30	76.81	—	20.94
	2	29.75	64.80	29.18	33.78	38.55	44.79	73.68	77.72	36.32	42.30	87.65	—	17.11
Canada	1	14.19	56.14	31.89	22.76	27.86	45.67	46.56	68.43	41.24	55.79	79.79	20.94	—
	2	13.84	79.02	28.87	21.85	28.65	42.18	41.17	61.77	36.09	56.86	64.02	17.11	—

Notes:

The distance indicator is based on the per cent distribution across 41 SIC classes of patents granted in the US to individual countries. The index varies from 0 to 1,000.

Distance values are divided by the maximum possible distance, calculated on the average of the five more extreme cases. Period 1: 1975–81; Period 2: 1982–8.

Source: CNR-ISRDS elaboration on CHI Research data.

tries, including the US, Japan, France, the UK and Germany, appear rather close, as they distribute their patents across all sectors.[9] Strong differences emerge, on the other hand, for small countries, which concentrate their patents in few (and different) areas.

Smaller countries are very different from one another and tend to be closer to the larger country which shares the same sectoral specialisation. Smaller countries, in other words, appear to have developed a technological specialisation in selected, country-specific 'niches' (see Walsh, 1988). Such a position allows small countries to compete in these fields at a world-class level. Their niches, however, are very different, showing that small nations can be specialised in entirely different areas: it is not just their size that determines the nature of their sectoral strengths and weaknesses.

The impact of selected sectors on national specialisation is also evident from the distance matrix of table 6.5. The strong similarity between France and the UK, and to a lesser extent the US, is due to the common importance of military-related and state-supported areas (aircraft, guided missiles, ordnance) and of classes such as drugs and medicines and agricultural and other chemicals. Some of these fields are areas of technological specialisation also for Germany, which, on the other hand, shares with Italy and Switzerland a relative concentration of its patenting in specialised industrial machinery and in some chemical classes.

Other types of links among national specialisations can be identified. A relative importance of electronics-related fields appears to be at the root of the similarities shown by Japan, the US, France and the Netherlands. The US is also fairly close to Canada, which in turn shows a specialisation profile similar to that of Sweden, with Denmark not too far apart; in these linkages the relative importance in all countries of natural resources and agriculture-related activities, as well as of specialised production in fields such as shipbuilding and engineering, appears to be crucial.

The analysis of the changes over time in the distance indicators is of particular interest as it highlights the different paths of specialisation shown by groups of countries. From the late 1970s to the mid-1980s, the US, the UK and France have further increased their already strong similarity. Germany, on the other hand moved away from this group, even though it remains close. Germany continues to be highly similar to Italy and is now closer to Switzerland and Sweden. In turn, Italy moved closer to the technological specialisation of Switzerland and Sweden, while increasing its distance from most other countries.

Japan is the clearest case of a country developing a distinct profile of specialisation in electronic and mechanical technologies, which is highly and increasingly different from all other countries. For the remaining countries (with the partial exception of the Netherlands) the general trend is towards a growing distance, suggesting that most countries develop a sectoral distribution of their patenting activities leading to a more distinct national profile, linked to their own economic and technological characteristics, rather than converging to a standard patenting profile for all advanced countries.

From this complex picture of relative similarities and differences a number of lessons can be drawn. First, we can argue that the convergence in terms of technological activities among the US, the UK and France, on the one hand, and among Germany, Switzerland, Sweden and Italy on the other hand, is rooted in two different and diverging patterns of sectoral specialisation. Japan shows an additional distinct pattern of technological activities which has led to the fastest catching up path among OECD countries. All these three different patterns of sectoral specialisation have been successful in leading most countries in these groups to converge at the top of the economic and technological performances of advanced countries.

Secondly, the ability of large countries to cover most technology fields with their innovative activities means that size (i.e., the aggregate volume of their resources devoted to innovation) is an important factor in this measure of similarity. This leads to a picture of a core made up of the major countries, which are fairly close to one another, while smaller ones appear scattered around them, closer to one of the larger countries, and highly different from most of the other smaller ones.

Thirdly, while the analysis of individual countries has shown strong stability in national patterns of specialisation in technology (see Cantwell, 1989), in comparing the relative distance between countries some mobility can be found. The differentiated pattern of sectoral patenting and the combined shifts over time of all countries led to an overall picture of fairly dynamic relative positions of individual countries. In other words, even within the constraints of the technological capabilities accumulated by each nation over time, countries do shift their relative positions, taking or missing the technological opportunities offered by the changing patterns of innovation.

Conclusions

In this chapter we have examined the technological basis of the process of economic convergence among advanced countries, showing that a moderate convergence can be found for indicators of aggregate technological efforts (see also Soete and Verspagen, 1993a). Our analysis for the restricted sample of OECD countries, however, indicates that 'technology gaps' remain much wider than 'economic gaps'. From a dynamic perspective, there is no evidence that convergence in technology is occurring at a higher pace than for levels of income. This result is consistent across different indicators of innovative activities. A significant divergence has emerged in the industrial R&D of the G5 group, where the US, the UK and France have not been able to keep up with the growth of resources in Japan and Germany.

More importantly, we have related the problem of convergence at the aggregate level to the sectoral specialisation of advanced countries, showing that diverging patterns of sectoral innovative activities can be identified even for the countries at the forefront of the 'convergence club'. This evidence supports the relevance of national systems of innovation for the understanding of how countries develop and use technology. The qualitative and institutional differences among countries highlighted by a new body of research (see, among others, Freeman and Lundvall, eds., 1988; Lundvall, ed., 1992, Nelson, ed., 1993) indicate the crucial role played by nation-specific factors and institutions in shaping both the patterns of technological change and countries' comparative advantages. The latter represent a major constraint also for the strategies of firms which have been increasing the internationalisation of technology. Cross-border strategies of multinational firms however do not necessarily bring about a greater uniformity among countries in the production of innovations; rather, they are more likely to lead to an increasing diversity (see the essays in Archibugi and Michie, eds., 1997).

Complementing the qualitative analyses carried out so far on the national systems of innovation, this chapter provides quantitative evidence on the similarities and differences among advanced countries in their economic and technological performances, at both the aggregate and the sectoral level. We have shown the evidence and the limitations of the process of convergence in technology and the different sectoral patterns of specialisation which may sustain it. Each country appears to have further developed its comparative advantages in selected technological niches.[10]

However, the increasing specialisation in the production of innovations does not necessarily mean that national patterns in the adoption and diffusion of technology, as well as patterns of market demand, are equally diverging (see Eaton and Kortum, 1995). In particular, the aggregate picture of the technological and economic performance of advanced countries provided above shows that a process of aggregate convergence is taking place. This apparent paradox is related to the ability of national economies to use their different technological resources, and to acquire the know-how they do not produce themselves from a variety of other channels, including trade, patent licensing, international cooperative research projects, and many other specific forms of technology transfer.

An interesting case of widely differing national systems of innovation, national patterns of sectoral specialisation in technology and moderate convergence is offered by EC countries (see Grupp, in this book). One may even suggest that the diversity among them should be regarded as an advantage, since it may make it possible to develop technological capabilities in different fields. However, in terms of developing a distinct profile of technological specialisation, Europe still appears as an aggregate of disparate countries – some of them rather similar (the UK and France, or to a lesser extent Germany and Italy), others more different (all the smaller ones) – and neither a common specialisation profile nor a clear complementarity is evident. The pattern of growing technological distance among EC countries does not appear as a clear outcome of either a policy of European integration or of a market-driven process of greater intra-European division of labour.

However, the most serious problem for European countries remains the incomplete process of convergence. In most indicators of technological activities dramatic lags persist for many countries which have shown strong economic convergence but little ability in research and innovation. Although effective use, adaption and diffusion of foreign-developed technology has helped the convergence in per-capita incomes, one may wonder whether there is room left for additional progress without a serious increase in domestic innovative activities, reducing the large gaps that are still visible for many countries. It should be clear however that any expansion of national aggregate innovative efforts is likely to take each country in different directions at the sectoral level, as shown by the patterns of specialisations discussed above.

In conclusion, the evidence on technological activities of advanced countries shows that in the last two decades economic convergence has been achieved with relatively little increase in technological efforts. In the 1980s the convergence process has slowed down, perhaps also as a consequence of strong persisting differences in countries' innovative activities. When we look beyond the aggregate picture, and consider the sectoral patterns of technological specialisation, we are left with a baffling paradox: countries converge by becoming more different and grow by becoming more specialised.

Notes

1 See, most recently, Abramovitz (1986), Baumol et al. (1989), Arrighi (1991), Maddison (1982, 1991), Dollar and Wolff (1993), Dosi and Fabiani (1994), Verspagen (1994).
2 See, among others, Hughes (1986) and Fagerberg (1988) for international competitiveness; Baumol *et al.* (1989) for productivity; Fagerberg (1987), Amable

(1993b) and Soete and Verspagen (1993b) for growth rates; Beelen and Verspagen (1994) for trade. An overview is provided in OECD (1992a).

3 On national systems of innovations see in particular Nelson, ed. (1993) and Lundvall, ed. (1992). See also Porter (1990).

4 The impact of different degrees of technological specialisation on countries' growth is investigated in Pianta and Meliciani (1996).

5 The decline of the US technological leadership in many industries is addressed by a large literature: see Pianta (1988) and Scherer (1992).

6 The same analysis has been carried out also considering patents per capita, obtaining similar results. However, this indicator overestimates the innovative activity of small countries with a high international integration.

7 A description of national patterns of sectoral specialisation, based on a variety of patent data and other indicators is provided in Casson, ed. (1991), Freeman et al., eds. (1991) and Archibugi and Pianta (1992).

8 Since the yearly number of patents is biased by fluctuations, we have grouped patenting for a fairly large number of years.

9 In previous research (Archibugi and Pianta, 1992) we have shown that the degree of specialisation in technology is inversely related to the size of technological activities of each country.

10 The increasing degree of technological specialisation (measured using patent data) of OECD countries and the positive impact it has had on growth, catching up and convergence is shown in Pianta and Meliciani (1996).

7 International patterns of technological accumulation and trade

GIOVANNI AMENDOLA, PAOLO GUERRIERI AND PIER CARLO PADOAN

1 Introduction

During the eighties the notion that technology plays a key role in explaining trade performance was supported by a number of empirical studies (Soete, 1987; Fagerberg, 1988a; Dosi, Pavitt and Soete, 1990). The strong linkage between technological innovation and international competitiveness found at the aggregate level (Fagerberg, 1988a) has also been pointed out at the sectoral level, albeit with notable differentiations (Soete, 1987; Dosi, Pavitt and Soete, 1990).

On the whole, the empirical evidence points to the linkage between a country's technological capacity and its ability to penetrate foreign markets. There exists, however, only a limited analysis of the evolution of technological and trade specialisation patterns, as well as of the changing relation between the two. In the following we will address both issues.

The chapter is structured as follows. Section 2 offers a brief overview of the literature on technological and trade specialisation. Section 3 deals with the serious problems of availability and standardisation of data for international comparisons, and describes the indicators we have used. In section 4 the concentration of national patterns of specialisation is discussed. Section 5 focusses on the persistence over time of these patterns. In section 6, the impact of technological specialisation on trade specialisation is assessed. A final section draws some conclusions.

2 Trade specialisation and technological specialisation: an overview of the literature

Few studies have focussed on the evolution of national patterns of trade specialisation. International economists have dedicated far more time and energy to the study of the determinants of trade patterns than to their

141

evolution. Nevertheless, a trend towards convergence in the sectorial specialisation of countries has been identified mainly through the analysis of intra-industry trade.[1] The expansion of intra-industry trade has led to a greater similarity in the export vectors of industrialised countries at the macro-sectorial level. At the same time, there also seems to be a trend towards product specialisation within the same sectors.[2]

On the whole, however, empirical studies on the evolution of trade specialisation have not reached robust conclusions as far as the convergence hypothesis is concerned, since the latter is heavily dependent upon the degree of commodity disaggregation chosen for analysis.

The literature available on the evolution of technological specialisation is even scanter. Nevertheless, the findings of Pavitt (1988a) show that the differentiated, firm-specific and cumulative nature of technology, emphasised in the literature on innovation (Atkinson and Stiglitz, 1969; Rosenberg, 1976 and 1982; Sahal, 1981) affects the technological development of firms as well as of nations. In addition, as the composition of the technological activities of a country is the sum of the firms there located, it tends to evolve gradually and cumulatively (Pavitt, 1988a).

Pavitt's results support his theory of technological accumulation. The comparison of the vectors of technological advantage in two periods (1963–8 and 1976–81) confirms the stability of the patterns of technological specialisation in nine out of ten major industrialised countries. The theory of technological accumulation has found further empirical support in the tests carried out by Cantwell (1989). Technological advantages have been stable throughout the sixties and seventies; while a test relative to 1890–2 and 1910–12 shows that they were stable during those periods as well. It also emerges that when considering the evolution of technological advantages over longer time horizons (1892–1912 and 1963–83), eight out of ten countries display a positive – though weak – linkage between the composition of innovative activities at the beginning and at the end of the periods examined (Cantwell, 1989).

The findings of Pavitt and Cantwell point to a remarkable stability in the international patterns of technological specialisation in the short and medium term (15–20 years). This result has a microeconomic basis, since it is consistent with the substantial stability in the short and medium term both of the characteristics of technology and the patterns of technological specialisation of firms. However, over longer periods (e.g., 50 years), the persistence of the international patterns of technological specialisation tends to fade away. The emergence of new technological paradigms and of new industries generates abrupt changes in firms' technological trajectories. Therefore, in the long term, substantial modifications in the profile of a country's technological specialisation are to be expected.[3]

A related point is the relation between the evolution of technological specialisation and the evolution of trade specialisation. Available empirical evidence does not provide a definitive and exhaustive answer. To date, this relation has been investigated only for selected periods. This is the approach used by Soete (1987) among others, who has calculated the rank correlation (the Spearman correlation coefficient) between the structure of technological advantages and the structure of trade advantages for OECD countries.[4] His results confirm the theoretical expectation of a positive relation between the two specialisations, even though the relation is only significant for smaller countries.

Further evidence is provided by studies which have considered other factors – in addition to technology – such as capital per head and the intensity of skilled labour in explaining trade patterns. However, this strand of analysis also has serious limitations. Most of these studies usually consider one country at a time, for given years, and are based on non uniform sectorial classifications. As a consequence, the possibility of international comparisons is often seriously limited.

The evolution of national patterns of specialisation, and the relation between technological specialisation and trade specialisation, for single countries over time, have not been thoroughly explored so far. The aim of this chapter is to start to fill this gap.

3 Problems of classification and methodology

In the previous section, we stressed that the assessment of trends in specialisation/despecialisation in international trade depends to a large degree on the level of commodity disaggregation. Similarly, we cannot assume that patterns of technological specialisation evolve differently from patterns of trade unless comparison is based on a uniform sectorial classification. In order to analyse the evolution of the two kinds of specialisation, as well as the changing relationship between them, a common reference classification must be set up.

We assume that the technological activities of each country can be adequately measured by its patenting in the United States. The choice of US patenting as a technology output indicator follows a tradition of empirical analysis (Soete, 1987; Pavitt, 1988a; Cantwell, 1989). An indicator of innovativeness based on US patenting is marked by some drawbacks (see the critical survey by Pavitt, 1988b), however it allows a uniform comparison – in the most advanced technological market – of the innovative capacities of different countries.

Furthermore, patents granted in the USA are particularly well suited for the investigation of the impact of technical change on trade performances

at the sectorial level. In the United States patents are classified according
to the US Patent Classification, however US patent data are also dissemi-
nated according to the US Standard Industrial Classification.[5] This allows,
among other things, to establish a correspondence with the classification
adopted for international trade data.

To this end, we have established a correspondence between the US
Standard Industrial Classification and the Standard International Trade
Classification, the reference nomenclature for international trade
adopted by the UN and the OECD. Details on data sources as well as
on the concordance adopted are given in the appendix 1. The harmonisation procedure has allowed the setting up of a database for 38 sectors
(listed in the appendix 2), for a relevant period of time (from 1970 to
1987 for foreign trade, and from 1967 to 1987 for patenting in the United
States).

Before discussing the empirical evidence, the following methodological
remarks are in order.

1 The 38 sectors considered include food products, oil, natural gas and oil
   products. The technological and trade specialisations of the major industrial countries will, therefore, be assessed with reference to a product set
   that is broader than manufacturing industry.
2 The reference classification reflects the shares of industrial sectors in
   terms of patenting; only those sectors in which patenting is intense are
   presented separately. As a result, traditional sectors, that account for a
   low share of US patenting, are condensed into three sectors: textile products; ceramics, glass and building materials; other manufactures. This
   last aggregation includes sectors such as wood and wood products, fur-
   niture, skins and leather, and footwear and luggage, which are relevant
   especially for Italian export. This point should be borne in mind since,
   on the basis of the classification adopted, Italian trade specialisation
   takes on a specific configuration which can be traced back partly to the
   fact that the sectors in which Italy is specialised are condensed into the
   residual grouping 'other manufactures'.
3 The technological specialisation profile of the United States reflects the
   influence of domestic patenting. Domestic patents are not, however, a
   reliable indicator, since they include less relevant patents (often produced
   by individual inventors) that do not require protection of patenting
   abroad. As a consequence, domestic patents do not offer a clear indication of a country's pattern of technological specialisation. This conclusion is especially important with respect to the United States, whose
   specialisation profile as measured by domestic patenting is only weakly
   correlated to the specialisation profile emerging from foreign patenting

(Archibugi and Pianta, 1990). Due to this limitation, we will not stress the results we have obtained for the United States.

As a next step, indicators for trade and technological specialisation have been computed for ten countries. The countries considered are: the United States, Japan, Germany, France, the United Kingdom, Italy, Canada, the Netherlands, Sweden and Switzerland.

Trade specialisation has been measured by two indicators. The first one is the well-known Balassa (1965) specialisation index. This index (henceforth referred to as RCA – Revealed Comparative Advantages indicator) takes on values greater than 1 when a country's market share in a given sector is greater than its share in total trade; it takes on values between 0 and 1 when the sectorial market share is below the average share. Values greater than 1 reveal trade specialisation; values less than 1 point to relative despecialisation.[6]

The second trade specialisation indicator is the Contribution to Trade Balance (CTB) initially introduced by CEPII (1983). Unlike RCA (which is based only on exports) this indicator also takes imports into consideration in measuring a country's specialisation. Values greater than 0 (less than 0) of the CTB indicator identify those sectors which give a contribution to the overall trade balance higher (lower) than their percentage share in the country's total trade, thus revealing sectors of specialisation (despecialisation) (Guerrieri and Milana, 1990).[7]

The Balassa index has been also used for technological specialisation after appropriate adaptation (henceforth referred to as RTCA – Revealed Technological Comparative Advantages indicator). In practice, instead of a country's share of world exports, its share of total patents granted in the United States is considered.

The indicators chosen were calculated for the periods 1970–3, 1977–80 and 1984–7 for trade specialisation and for longer periods – 1967–73, 1974–80 and 1981–7 – for technological specialisation in the attempt to smooth the intrinsically erratic nature of patent data.

4 The concentration of national patterns of specialisation

The degree of concentration of each national pattern of specialisation is the first issue to be addressed. To this end, we have measured the variance (dispersion) of the RCA and RTCA indicators with respect to their mean (theoretical) value, which is equal to 1 (see table 7.1). High values of the variance reveal strong specialisation in some sectors and strong despecialisation in others. These can be referred to as 'concentrated' patterns of specialisation. Low values of the variance show the opposite case, which can be defined as one of 'diffused' specialisation.

Table 7.1. *The degree of technological specialisation and trade specialisation**

Countries	Indicator of technological specialisation – RTCA (Revealed Technological Comparative Advantages)			Indicator of trade specialisation RCA – (Revealed Comparative Advantages)		
	1967–73	1974–80	1981–7	1970–3	1977–80	1984–7
Canada	0.229	0.190	0.257	0.585	0.708	0.812
France	0.167	0.166	0.057	0.352	0.382	0.566
Germany	0.094	0.103	0.143	0.367	0.424	0.342
Japan	0.216	0.198	0.204	1.764	2.020	1.018
Italy	0.300	0.256	0.286	0.430	0.351	0.385
Netherlands	0.244	0.332	0.294	0.817	0.769	0.767
United Kingdom	0.142	0.111	0.147	0.261	0.281	0.211
United States	0.007	0.012	0.019	0.734	0.756	0.840
Sweden	0.216	0.182	0.336	0.433	0.884	0.721
Switzerland	0.491	0.362	0.376	2.596	2.611	2.109

Note:
* Variance of the indicators with respect to their mean.
Source: Trade data from SIE World Trade Data Base, which is based on the OECD and UN trade statistics; Patent data from the US Patent and Trademark Office.

The results are highly suggestive. First, in each country the pattern of trade specialisation appears to be more concentrated than the equivalent pattern of technological specialisation. If foreign trade is taken as a proxy of a country's production, then the result at the aggregate level is consistent with the empirical evidence at the micro level indicating that technological diversification of firms is much higher than their product diversification (Pavitt et al., 1989).[8] We assume that the diversity between the concentration of technological specialisation and the concentration of trade specialisation, which we find at the country level, stems from a corresponding asymmetry at the firm level.[9]

Table 7.1 also offers interesting insights on individual countries characteristics. In all of the periods considered, Switzerland is the country showing the highest specialisation, both in technology and in trade. Sweden and the Netherlands are also highly specialised. This suggests that the smaller countries we have examined display a relatively high level of specialisation reflecting the fact that the need to reach minimal dimensions in both R&D programmes and in production forces smaller countries to specialise in specific sectors.[10] As regards other countries, specialisation is very high in Japan, but only in the foreign trade component. The United Kingdom stands out for its low degree of specialisation, both in technological activities and in exports.[11]

Italy stands out, among the seven major industrialised countries, for its relatively high technological specialisation. In the three periods considered, Italy ranks second, third and fourth, respectively (in the final period it ranks behind Switzerland, Sweden and the Netherlands only). This somewhat surprising result can be explained by the strong dichotomy that marks the technological specialisation of the Italian industry at the sectorial level: on the one hand, technological specialisation is very high in sectors such as household appliances, agricultural chemicals, pharmaceutical products and specialised industrial machinery; on the other hand, in some sectors such as miscellaneous chemical products, electrical apparatus and materials, and in a number of electronic sectors, the Italian industry displays a strong technological despecialisation.

The relative concentration of Italy's pattern of specialisation emerges with respect to technological activities only, however. The variance of the RCA indicators does not differ significantly from the values obtained for France and Germany, although one should bear in mind that the reference classification we have adopted leads to an underestimation of the trade specialisation of the Italian industry for the reasons mentioned in section 3.

Table 7.1 reveals that not only Italy, but her three major Community partners show low levels of trade specialisation: in the three periods considered, France, Italy, Germany and the United Kingdom stand at the bottom of the rank. Increasing intra-industry trade among European Community countries can be an important factor in explaining such a flattening of the trade specialisation profiles of the major Community countries.

Finally, no significant regularities can be found in the evolution of the variance of the RCA and RTCA indicators: specialisation increases in some countries (Canada, the United States and Sweden), while it decreases in others (Japan, Italy and Switzerland). Others, such as France, Germany, the Netherlands and the United Kingdom, show opposite trends in the two specialisations.

One final remark is in order: the empirical evidence based on the sectorial disaggregation we have adopted (38 sectors) does not support the hypothesis that foreign trade evolves towards sectorial despecialisation, since the specialisation in exports of four out of ten countries (Canada, France, the United States and Sweden) has increased, at least in the 1970–87 period. Moreover, careful inspection of the six countries which show a trend towards despecialisation reveals that the degree of trade specialisation fell significantly only in the cases of Japan and Switzerland.

5 The cumulative nature of national patterns of specialisation

In this section the hypothesis that national patterns of specialisation evolve gradually and cumulatively will be tested. For this purpose, the sectoral distributions of each specialisation indicator have been compared in two different periods, estimating simple cross-section regressions.[12] More specifically, we have adopted the Galtonian regression model, a technique devised for the analysis of bivariate distributions. This statistical methodology has allowed for the checking of whether the specialisation vectors of each country are stable or whether they tend to change over time.

Formally, the relation estimated for each of the ten countries and for each of the three periods considered is as follows:

$$X_{it} = a + bX_{it-j} + u_{it} \qquad i = 1,\ldots,N; \ j = 1,2$$

where X refers to both the indicator for technological specialisation RTCA and the two indicators used to measure trade specialisation (RCA and CTB), where i indicates the sector.

Before discussing the results, we must clarify the interpretation of the regression coefficient. The hypothesis of perfect stability in the structure of a country's technological (or trade) advantages corresponds to a regression coefficient of 1. Values of b above 1 show greater specialisation of the country in sectors where revealed advantages already exist and a trend towards despecialisation in relatively disadvantaged sectors. Values of b between 0 and 1 indicate a weakening of specialisation in advantaged sectors and a strengthening of specialisation in disadvantaged sectors. In such a case, statistically defined as the 'regression towards the mean', the gap between the sectors initially advantaged and those initially disadvantaged decreases.[13] Finally, values of b below 0 denote that the ranking of the sectors has deeply changed, and even possibly inverted itself (the case of $b = -1$).

Regressions results (see tables 7.2–7.4) apparently confirm the remarkable stability over time of both the patterns of technological specialisation and the patterns of trade specialisation. The regression coefficients are very high, standing usually between 0.7 and 1. Values of b above 1, indicating a deepening of the specialisation profiles, are also rather frequent. Furthermore, t-statistics show that the theoretical hypothesis of constant patterns of specialisation over time (parameter $b = 1$) cannot be rejected in 53.3 per cent of the equations estimated on the RTCA and RCA sectorial distributions (the patterns of trade as measured by the CTB indicator are,

Table 7.2. *Regression results:* $RTCA_t = a + b\, RTCA_{t-j} + u_t$ $(j=1,2)$

(Period I: 1967–73; Period II: 1974–80; Period III: 1981–7)

Countries	P	b	R²	Countries	P	b	R²
	I–II	0.794 (−2.732)***	0.755		I–II	1.240 (−2.637)**	0.838
Canada	II–III	1.032 (0.365)	0.794	United States	II–III	1.122 (1.122)	0.778
	I–III	0.909 (−1.011)	0.734		I–III	1.426 (2.645)**	0.684
	I–II	0.609 (−2.973)***	0.374		I–II	0.396 (−4.820)***	0.218
France	II–III	0.405 (−8.451)***	0.479	United Kingdom	II–III	0.936 (−0.506)	0.603
	I–III	0.323 (−8.358)***	0.306		I–III	0.212 (−4.719)***	0.043
	I–II	0.893 (−1.180)	0.729		I–II	0.787 (−2.609)**	0.720
Germany	II–III	0.781 (−1.557)	0.461	Italy	II–III	0.886 (−1.196)	0.706
	I–III	0.733 (−1.682)	0.372		I–III	0.601 (−3.104)***	0.378
	I–II	0.813 (−2.172)**	0.712		I–II	0.974 (−0.242)	0.695
Japan	II–III	0.919 (−1.103)	0.813	Netherlands	II–III	0.872 (−2.177)**	0.859
	I–III	0.710 (−2.565)**	0.523		I–III	0.934 (−0.680)	0.723
	I–II	0.799 (−2.683)**	0.760		I–II	0.782 (−3.712)***	0.832
Sweden	II–III	1.154 (1.279)	0.719	Switzerland	II–III	0.968 (−0.568)	0.894
	I–III	0.976 (−0.185)	0.612		I–III	0.749 (−3.277)***	0.727

Notes:

t-statistics in parentheses refer to the hypothesis H_0: $b=1$.

*** significant at the 1 per cent level;

** significant at the 5 per cent level;

* significant at the 10 per cent level.

Table 7.3. *Regression results:* $RCA_t = a + b\,RCA_{t-j} + u_t$ $(j=1,2)$

(Period I: 1970–3; Period II: 1977–80; Period III: 1984–7)

Countries	P	b	R^2	Countries	P	b	R^2
	I–II	1.009 (0.093)	0.733		I–II	0.925 (−1.361)	0.887
Canada	II–III	1.024 (0.450)	0.909	United States	II–III	0.995 (−0.057)	0.802
	I–III	0.951 (−0.351)	0.564		I–III	0.973 (−0.322)	0.796
	I–II	0.869 (−1.878)*	0.811		I–II	0.888 (−1.344)	0.759
France	II–III	1.197 (2.633)***	0.876	United Kingdom	II–III	0.763 (−2.514)**	0.646
	I–III	0.970 (−0.234)***	0.618		I–III	0.579 (−3.258)***	0.358
	I–II	0.961 (−0.772)	0.910		I–II	0.715 (−3.150)***	0.635
Germany	II–III	0.894 (−1.608)	0.836	Italy	II–III	0.986 (−0.213)	0.870
	I–III	0.820 (−1.983)*	0.692		I–III	0.632 (−3.117)***	0.443
	I–II	1.005 (0.117)	0.929		I–II	0.931 (−1.640)	0.932
Japan	II–III	0.690 (−7.718)***	0.891	Nether- lands	II–III	0.957 (−0.796)	0.895
	I–III	0.655 (−5.294)***	0.737		I–III	0.907 (−1.555)	0.866
	I–II	1.185 (1.522)	0.726		I–II	0.955 (−1.184)	0.947
Sweden	II–III	0.859 (−2.616)**	0.877	Switzer- land	II–III	0.892 (−4.418)***	0.973
	I–III	0.893 (−0.705)	0.489		I–III	0.841 (−3.385)***	0.900

Notes:

t-statistics in parentheses refer to the hypothesis H_0: $b=1$.

*** significant at the 1 per cent level;

** significant at the 5 per cent level;

* significant at the 10 per cent level.

Table 7.4. *Regression results:* $CTB_t = a + b\,CTB_{t-j} + ut$ $(j=1,2)$

(Period I: 1970–3; Period II: 1977–80; Period III: 1984–7)

Countries	P	b	R²	Countries	P	b	R²
	I–II	0.621 (−10.938)***	0.899		I–II	0.890 (−0.513)	0.325
Canada	II–III	0.408 (−5.759)***	0.304	United States	II–III	0.431 (−13.503)***	0.744
	I–III	0.195 (−10.907)***	0.162		I–III	0.564 (−4.853)	0.523
	I–II	1.457 (5.642)***	0.900		I–II	0.415 (−19.442)***	0.841
France	II–III	0.779 (−7.568)***	0.952	United Kingdom	II–III	0.158 (−3.494)***	0.012
	I–III	1.111 (1.289)	0.822		I–III	−0.021 (−9.316)***	0.001
	I–II	0.959 (−0.452)	0.756		I–II	1.343 (3.298)***	0.823
Germany	II–III	0.868 (−4.513)***	0.961	Italy	II–III	0.927 (−3.474)***	0.982
	I–III	0.873 (1.734)*	0.799		I–III	1.227 (2.115)**	0.784
	I–II	1.424 (3.614)***	0.804		I–II	0.897 (−1.892)*	0.883
Japan	II–III	0.773 (−6.933)***	0.939	Netherlands	II–III	0.902 (−3.111)**	0.958
	I–III	1.030 (0.243)	0.661		I–III	0.811 (−3.323)***	0.850
	I–II	1.008 (0.116)	0.855		I–II	0.899 (−1.889)	0.887
Sweden	II–III	0.647 (−14.396)***	0.951	Switzerland	II–III	0.963 (−1.074)	0.957
	I–III	0.657 (−6.806)***	0.825		I–III	0.864 (−2.200)**	0.844

Notes:
t-statistics in parentheses refer to the hypothesis H_0: $b=1$.
*** significant at the 1 per cent level;
** significant at the 5 per cent level;
* significant at the 10 per cent level.

instead, rather unstable). Finally, in most of the estimated equations the coefficient R^2 is very high: the evolution of national patterns of specialisation can thus largely be explained by countries' initial specialisation profiles.

Comparison of tables 7.2 and 7.3 shows that, for contiguous periods, regression coefficients relative to trade specialisation indicators are generally higher than those relative to technological specialisation indicators. This result (which would require a more careful scrutiny) apparently suggests that cumulative trends in trade specialisation are stronger than those prevailing in technological specialisation. Such a finding – i.e., the cumulative nature of trade specialisation – would lend support to the model of trade suggested by Krugman (1987), in which arbitrary patterns of specialisation, once established, tend to persist and extend over time.

Examination of the regressions with variables relative to the two most distant periods (the first and the third) confirms a remarkable stability both in the patterns of technological specialisation and in the patterns of trade specialisation. Yet, in accordance with theoretical expectations – since the time interval is longer – in most cases the value of b decreases, denoting a weakening of the linkage between the initial and final specialisation structures. As we have already stressed, the emergence of new technological paradigms and new industries can, in the long term, generate important changes in the specialisation trajectories of both firms and countries. In fact, for each of the countries examined, there are sectors in which technological and trade advantages (disadvantages) have remained stable and sectors in which comparative advantages (disadvantages) have changed substantially during the longer time span considered (1967–87 for patenting and 1970–87 for trade).

With respect to individual countries, clear-cut trends marked by increasing technological specialisation can be found for the United States (the values of b are always greater than 1), Sweden, the Netherlands and Canada (see table 7.2). The most remarkable changes in technological specialisation profiles are found for France and the United Kingdom. In the case of France, technological advantages are constant only in the sectors dominated by public procurements and R&D financing (for example, aerospace, railroad vehicles and some sectors of a chemical 'filière' such as pharmaceuticals).[14] As for Great Britain, Cantwell (1989) has suggested a more general interpretation: the structure of the country's technological advantages can radically change when the technologically stronger firms decline and drop out of the market as a consequence of a decline in the rate of growth of productivity and innovative processes.

Let us now consider the evolution of trade specialisation. Regressions of the RCA indicators (between the first and the third period) confirm (see table 7.3) that the United States is the country that shows the strongest tendency towards a cumulative pattern of specialisation. France, the Netherlands and Sweden follow, displaying large coefficients. Japan, Italy and the United Kingdom are last in rank.

Nevertheless, in the case of Japan, the value of R^2 is very high – much higher than the value of b. This means that the reduction in the degree of specialisation induced by the 'regression effect' ($0 < b < 1$) has not been compensated for by the 'mobility effect' in the ranking of the sectors (see note 13). Sectorial analysis shows that Japan has not significantly changed its trade specialisation profile. While displaying a substantially stable sectorial hierarchy, the pattern of specialisation of Japanese industry has become more 'compact'; that is, it shows a lower variability among advantaged sectors and disadvantaged sectors (see section 4). The low values of R^2 in the cases of Italy and of the United Kingdom reveal remarkable changes in the ranking of the sectors. Both the pattern of technological specialisation and the export structure of these countries have changed significantly.

However, the evolution of trade specialisation, as measured by the CTB indicator (see table 7.4), shows significant differences. Italy has the most stable structure. France follows, confirming its status of a country with a very stable pattern of trade specialisation, while Japan comes next. The regressions coefficients of these countries between the first and third periods are greater than 1. Therefore, the initial structure of trade balances in Italy, France and Japan has further consolidated.

It should also be noted that for both Italy and Canada the evolution of trade specialisation in the period 1970–3/1984–7 varies according to the indicator considered. In particular, while the structure of Italian exports has changed to some extent as shown by the RCA indicators, the structure of the trade balance has substantially remained the same. The opposite holds for Canada.

Finally, once again the United Kingdom stands out with respect to all the other major countries, displaying a negative regression coefficient between the first and the third period. Our empirical evidence suggests, therefore, that over the 1970s and the 1980s, Britain's pattern of specialisation for both technology and trade has changed profoundly.

6 Technological specialisation as a determinant of trade specialisation

In this section the hypothesis that trade specialisation is determined by technological specialisation will be tested. Such a relationship is consistent with 'technology-gap' models of trade (Posner, 1961; Hufbauer, 1966) and especially with models within a comparative advantage framework (Krugman, 1990).

The hypothesis to be tested is formalised by regressing the two sets of trade specialisation indicators (RCA and CTB) on the technological specialisation indicator (RTCA). The independent variable is considered both simultaneously and lagged, consequently four groups of cross-section equations (1)–(4) have been estimated:

$$RCA = a + b \, RTCA + u_1 \tag{1}$$

$$RCA = a + b \, RTCA_{-1} + u_2 \tag{2}$$

$$CTB = a + b \, RTCA + u_3 \tag{3}$$

$$CTB = a + b \, RTCA_{-1} + u_4 \tag{4}$$

The results of the regressions (see tables 7.5–7.8) are quite encouraging for a 'technology-gap' account of trade flows. As expected, the sign of the regression coefficient is always positive (the only exceptions are given by two equations estimated for Canada). More importantly, the parameter b is significant at least at the 10 per cent level in 67 per cent of the estimated equations.[15] There exist, however, notable differences among the countries examined. A rapid inspection suggests a division into three groups of countries:

1 leaders in world trade (Germany and Japan), for which the coefficients are generally statistically significant in practically all cases considered;
2 intermediate-size countries (Canada, France, Italy and the United Kingdom) for which the coefficients are significant only in some periods;
3 small countries with significant coefficients (the Netherlands, Sweden, Switzerland).

The United States have been omitted from our discussion, as the results of the regressions are largely unreliable, confirming the bias in the patent indicators adopted for this country.

In addition to the above initial grouping, relevant national specificities emerge within each group of countries. For Germany, b is significant in equation (1) only in the first two of the three periods considered, while it is significant in both cases of equation (2). Equations (3) and (4) display more significant results, although the absolute value of b tends to decrease for the more recent periods.

The characteristics of the German case are more evident when compared with those of Japan. Parameter b, as estimated in equation (1) is highly significant in all three periods for Japan; moreover, its absolute value is higher than the German value. The estimate of b obtained from equation (2) is relatively less significant, as are the estimates of b from (3) and (4); all these results are in contrast with those obtained for Germany.

The apparently weaker effect of RTCA on CTB in Japan as compared with the case of Germany can be explained by taking into account that Japan's imports are relatively contained both in sectors of relative strength and in disadvantaged sectors. Positive contributions to the trade balance show high values and are concentrated in a very small number of sectors (automobiles, motorcycles and electronics). As a result, Japanese technological specialisation provides a better explanation for trade specialisation in foreign markets (exports) than in the domestic market (imports). In Germany, technological specialisation instead seems to explain trade specialisation on the domestic market as well.

Furthermore, the above findings indicate significant differences between the two countries, as regard the evolution of technological and trade specialisation over the 1980s (Amendola, Guerrieri and Padoan, 1992; Guerrieri, 1992). Japan's experience shows a strengthening in the linkage between the two patterns, thanks to a positive trade performance and an increasing technological specialisation in sectors such as the electronic industries; whereas Germany's trade specialisation has increased in specific sectors (especially traditional products) and has decreased in others (high-tech industries), displaying a different trend with respect to the evolution of technological specialisation and, hence, a weakening linkage between the two patterns over the last decade.

Relevant national characteristics also emerge within the second group of countries. In the case of France the significance of the coefficient increases over time. Parameter b, in the case of equation (1), becomes significant in the more recent periods and takes on a high absolute value in the third period. A similar though weaker trend emerges from the results of the estimates of equation (2). The estimates of equation (3), but not those of equation (4), confirm this trend. The results of the estimates for France seem to indicate a case of a changing pattern of technological specialisation, which tends to become more similar to that of the stronger countries at the sectoral level (Amendola, Guerrieri and Padoan, 1992). However, such an interpretation must be considered prudently, given France's unsatisfactory trade performance over the 1980s. In this regard, two contrasting explanations could be considered: one would suggest that the strengthening of technological specialisation has not yet affected French trade performance because, as is well known, an adequate time span is requested to translate technological advantages into trade advantages; the other would suggest that the concentration of French technological advantages in sectors dominated by state-owned firms and public procurements is responsible for the difficulties of gaining trade advantages on world markets.

The results for Italy are easier to interpret. Parameter b is significant in equation (1) only in the first period. In general, trade specialisation

Table 7.5. *Regression results*: RCA $= a + b$ RTCA $+ u_1$

(RCA: I=1970–3; II=1977–80; III=1984–7; RTCA: I=1967–73; II=1974–80; III=1981–7)

Countries	P	a	b	R^2	Countries	P	a	b	R^2
Canada	I	0.178 (0.678)	0.485 (2.112)**	0.085	United Kingdom	I	0.912 (3.790)	0.278 (1.339)	0.048
	II	0.579 (1.705)*	0.241 (0.773)	0.016		II	0.875 (3.013)***	0.326 (1.311)	0.046
	III	0.918 (2.710)**	−0.094 (−0.322)	0.003		III	0.523 (2.516)**	0.528 (2.930)***	0.193
France	I	0.857 (3.183)***	0.303 (1.316)	0.046	Italy	I	0.487 (2.361)**	0.566 (3.235)***	0.225
	II	0.749 (3.018)***	0.497 (2.333)**	0.131		II	0.738 (3.587)***	0.357 (1.954)*	0.096
	III	−0.317 (−0.679)	1.517 (3.523)***	0.256		III	0.752 (3.497)***	0.295 (1.584)	0.065
Germany	I	0.583 (2.057)**	0.691 (2.573)**	0.155	Netherlands	I	0.425 (1.360)	0.721 (2.617)**	0.159
	II	0.631 (2.304)**	0.715 (2.805)***	0.179		II	0.504 (1.979)*	0.678 (3.071)***	0.207
	III	0.910 (3.395)***	0.314 (1.343)	0.049		III	0.677 (2.371)**	0.454 (1.775)*	0.080

Japan	I	−0.027	1.208	0.182
		(−0.056)	(2.832)***	
	II	0.262	1.114	0.132
		(0.488)	(2.342)**	
	III	0.212	0.968	0.194
		(0.589)	(2.943)***	
Sweden	I	0.174	0.874	0.378
		(0.871)	(4.681)***	
	II	0.584	0.658	0.093
		(1.625)	(1.926)*	
	III	0.951	0.167	0.013
		(3.334)***	(0.692)	
Switzerland	I	−0.165	1.566	0.476
		(−0.514)	(5.719)***	
	II	0.104	1.427	0.302
		(0.260)	(3.944)***	
	III	0.116	1.243	0.294
		(0.307)	(3.874)***	

Notes:

t-statistics in parentheses.

*** significant at the 1 per cent level;

** significant at the 5 per cent level;

* significant at the 10 per cent level.

Table 7.6. *Regression results:* $RCA = a + b\,RTCA_{-1} + u_2$

(I: RCA_{-1}=1977–80; $RTCA_{-1}$=1967–73; II: RCA=1984–7; $RTCA_{-1}$=1974–80)

Countries	P	a	b	R^2	Countries	P	a	b	R^2
Canada	I	0.362 (1.134)	0.440 (1.586)	0.965	United Kingdom	I	0.965 (3.917)***	0.251 (1.183)	0.037
	II	0.847 (2.284)**	−0.022 (−0.064)	0.001		II	0.848 (3.039)***	0.225 (0.941)	0.024
France	I	0.969 (3.735)***	0.297 (1.335)	0.047	Italy	I	0.828 (4.048)***	0.257 (1.484)	0.057
	II	0.644 (2.000)*	0.588 (2.137)**	0.113		II	0.752 (3.390)***	0.301 (1.525)	0.061
Germany	I	0.673 (2.350)**	0.684 (2.520)**	0.150	Netherlands	I	0.414 (1.396)	0.751 (2.874)***	0.187
	II	0.655 (2.368)**	0.582 (2.260)**	0.124		II	0.579 (2.149)**	0.541 (2.312)**	0.129
Japan	I	0.351 (0.654)	0.999 (2.160)**	0.115	Switzerland	I	0.036 (0.113)	1.504 (5.488)***	0.455
	II	0.454 (1.132)	0.700 (1.975)*	0.098		II	0.119 (0.334)	1.333 (4.139)***	0.322
Sweden	I	0.485 (1.494)	0.759 (2.501)**	0.148					
	II	0.649 (1.937)*	0.493 (1.546)	0.062					

Notes:

t-statistics in parentheses.

*** significant at the 1 per cent level;

** significant at the 5 per cent level;

* significant at the 10 per cent level.

Table 7.7. Regression results: CTB=a+b RTCA+u₃

(CTB: I=1970-3; II=1977-80; III=1984-7. RTCA: I=1967-73; II=1974-80; III=1981-7)

Countries	P	a	b	R^2	Countries	P	a	b	R^2
	I	−1.792 (−1.635)	1.718 (1.798)*	0.082		I	−3.674 (−2.492)**	3.334 (2.623)**	0.161
Canada	II	−1.697 (−2.372)**	1.697 (2.587)**	0.157	United Kingdom	II	−1.479 (−1.797)*	1.314 (1.864)*	0.088
	III	−0.427 (−0.815)	0.408 (0.904)	0.022		III	−0.062 (−0.060)	0.058 (0.065)	0.001
	I	−2.037 (−1.979)*	1.855 (2.104)**	0.109		I	−1.348 (−1.287)	1.286 (1.451)	0.055
France	II	−1.553 (−0.938)	1.415 (1.419)	0.027	Italy	II	−1.769 (−1.102)	1.758 (1.233)	0.041
	III	−4.170 (−2.078)**	3.936 (2.127)**	0.112		III	−1.636 (−1.122)	1.601 (1.266)	0.043
	I	−6.085 (−3.943)***	6.020 (4.120)***	0.320		I	−2.201 (−3.139)***	2.154 (3.484)***	0.252
Germany	II	−6.152 (−3.625)***	6.002 (3.798)***	0.286	Netherlands	II	−1.285 (−2.052)**	1.287 (2.370)**	0.135
	III	−4.369 (−2.982)***	4.017 (3.148)***	0.216		III	−1.372 (−2.353)**	1.404 (2.691)**	0.167

Table 7.7 (cont.)

Countries	P	a	b	R²
	I	-5.186 (-2.271)**	4.859 (2.474)**	0.145
Japan	II	-6.624 (-1.654)	6.033 (1.798)*	0.082
	III	-4.974 (-1.735)*	4.974 (1.904)*	0.092
	I	-3.327 (-2.678)**	3.456 (2.971)***	0.197
Sweden	II	-3.669 (-2.485)**	3.817 (2.716)**	0.170
	III	-1.200 (-1.392)	1.159 (1.596)	0.066

Countries	P	a	b	R²
	I	-2.527 (-2.895)***	2.705 (3.614)***	0.266
Switzerland	II	-2.382 (-2.448)**	2.543 (2.904)***	0.189
	III	-2.455 (-2.472)**	2.437 (2.893)***	0.189

Notes:

t-statistics in parentheses.

*** significant at the 1% level;

** significant at the 5% level;

* significant at the 10% level.

Table 7.8. *Regression results:* $CTB = a + b\,RTCA_{-1} + u_4$

(I: CTB=1977–80; $RTCA_{-1}$=1967–73. II: CTB=1984–7; $RTCA_{-1}$=1974–80)

Countries	P	a	b	R²
Canada	I	−1.022 (−1.410)	0.979 (1.550)	0.063
	II	−0.876 (−1.583)	0.876 (1.727)*	0.077
France	I	−3.207 (−2.036)**	2.921 (2.165)**	0.115
	II	1.399 (−1.062)	1.275 (1.129)	0.034
Germany	I	−6.570 (−3.822)***	6.500 (3.994)***	0.307
	II	−5.307 (−3.494)***	5.177 (3.660)***	0.271
Japan	I	−5.821 (−1.540)**	5.454 (1.677)	0.073
	II	−3.973 (−1.294)	3.827 (1.407)	0.052
Sweden	I	−3.581 (−2.636)**	3.719 (2.925)***	0.192
	II	−2.282 (−2.301)**	2.375 (2.517)**	0.149
United Kingdom	I	−1.367 (−1.991)*	1.242 (2.098)	0.109
	II	−0.062 (−0.050)	0.056 (0.052)	0.001
Italy	I	1.374 (−0.872)	1.309 (0.983)	0.026
	II	−1.675 (−1.116)	1.667 (1.250)	0.042
Netherlands	I	−1.966 (−2.878)***	1.925 (3.196)***	0.221
	II	−1.137 (−1.958)**	1.138 (2.259)**	0.124
Switzerland	I	−2.340 (−2.777)***	2.504 (3.465)***	0.250
	II	−2.323 (−2.417)**	2.479 (2.867)***	0.186

Notes:

t-statistics in parentheses.

*** significant at the 1 per cent level;

** significant at the 5 per cent level;

* significant at the 10 per cent level.

(regardless of its specification) does not seem to be explained by technological specialisation. In effect, technological specialisation in Italy seems to explain trade specialisation only in some chemical, engineering and traditional sectors, where the country shows a satisfactory trade performance (Archibugi and Santarelli, 1989; Amendola, Guerrieri and Padoan, 1992).

The results for the United Kingdom require a different interpretation, as b is significant only in very few cases and, especially, it loses its significance in the last two periods. The case of the United Kingdom would seem to be the opposite of that of France: the trade pattern is becoming ever more unrelated to the technological pattern. Actually, this finding points to an increasing weakness of the British industrial structure.

In all equations considered, estimates for Canada display not highly significant parameters. This confirms that Canada represents a case of a small economy with high trade specialisation not influenced by technological specialisation.

The results obtained for the last three countries – all small and highly specialised – are quite different from those for Canada. In the Netherlands, the influence of RTCA on RCA is almost always significant, especially when RTCA is lagged (equation (2)). The results obtained with CTB look more solid, confirming that this is a common feature for most of the European Community countries.

In the case of Sweden, the effect of RTCA on RCA tends to be less significant over time. A similar trend appears in the equations where CTB is the dependent variable even if, as for the other European countries, the results obtained with equations (3) and (4) are more solid.

Finally, in the case of Switzerland parameter b is always significant and is generally higher than for the other countries. Switzerland is a clear example of a small economy highly specialised both in trade and technology with a strong linkage between the two specialisations, and marked by a strong persistence in its sectoral pattern of specialisation.

7 Concluding remarks

Technological specialisation and trade specialisation involve a large number of issues that require much further investigation, particularly from an empirical viewpoint. While theoretical contributions to the relationship between technological innovation and trade patterns have flourished over the recent past, relatively little attention has been devoted to the related empirical aspects, undoubtedly because of the difficulties one has to face. This study has considered some selected topics, such as the persistence over time of trade and technological specialisations, and the role of technological specialisation in determining trade specialisation, and has offered a contribution to their empirical analysis.

A preliminary problem was the establishment of an empirical base to compare trade and technological specialisation at a sufficiently disaggregate level and over a sufficiently long period to allow for a thorough assessment of changing patterns of specialisation.

Our results indicate that the patterns of trade specialisation are more concentrated than those of technological specialisation. The hypothesis of increasing trade despecialisation, made popular by the literature on intra-industry trade, has not been confirmed. Moreover, international patterns of technological and, in particular, trade specialisation, show a remarkable stability over time, confirming the hypothesis recently put forward by Pavitt (1988a) as regards the former and by Krugman (1987) as regard the latter. It must be pointed out, though, that stability tends to weaken over longer periods of time.

With respect to the linkage between the two specialisations, the countries considered can be divided into three groups: leaders in technology and trade (Germany and Japan), with strong linkages between technological specialisation and trade specialisation; intermediate countries (France, the United Kingdom, Italy and Canada), with weaker linkages between the two specialisations, also as a result of their relatively rapid evolution; small countries (the Netherlands, Switzerland and Sweden), with persistent patterns of specialisation and strong linkages between trade and technological specialisation.

To sum up, despite the drawbacks of the chosen technological indicator, our findings corroborate a 'technology-gap' account of trade flows. The relationship between technological specialisation and trade specialisation, however, is strongly affected by significant country-specific factors, confirming that variety, not only among firms, but also among countries, is a distinctive feature of technical change.

Appendix 1

Trade data in current values (US dollars) are from SIE-World Trade database which is derived from the UN and OECD official trade statistics. Data on patents granted in the USA (fractional counts) are from the US Department of Commerce, Patent and Trademark Office.

As regards the concordance established, the difficulty of making comparable the US Standard Industrial Classification (SIC) and the Standard International Trade Classification (SITC) – two quite different systems of classification – has been particularly complicated by the time factor.

In fact, the analysis of the evolution of patterns of technological and trade specialisation requires an appropriate period of time and, hence, a correspondence between the SIC and the SITC that takes into account the changes in nomenclature introduced in 1978 after the second revision of the SITC. Such a revision has complicated disaggregate analysis of the evolution of international trade, and it has also discouraged the comparison of the evolution of technological and trade specialisation. For instance, the work by Soete (1987) – the most thorough to date on the relation between the two kinds of specialisations – takes 1977 as the reference year. As Soete puts it 'the analysis is purely static, all the evidence presented relates to one year (1977)' (1987, p.101).

To overcome these difficulties we have used the classification of 400 product groups of the SIE World Trade database (Guerrieri and Milana, 1990; Guerrieri, 1991). This classification is derived from the approximately, 1900 products considered in original UN and OECD statistical sources. Its main advantage lies in the fact that it allows for analysis of extremely deseg-regated time series, given its system of correspondence between the SITC Revised (in force from 1961 to 1977) and the SITC Revision 2 (used since 1978). Furthermore, the trade classification adopted here is compatible with the International Standard Industrial Classification (ISIC). This has allowed the setting up of a correspondence with the SIC, as the structure of the latter is similar to that of the ISIC. In particular, in establishing a correspondence between the SIC and the trade classification, we have kept the correspondence between the SIC and the SITC Revised proposed by Soete (1987); this, of course, has required modifications taking into account the introduction of the SITC Revision 2 in 1978. In many cases, these modifications seem to have improved the arrangement adopted by Soete.

Appendix 2 List of industries

NUMBER	PRODUCT FIELD	SIC CODE
1	Food and kindred products	20
2	Textile mill products	22
3	Industrial inorganic chemistry	281
4	Industrial organic chemistry	286
5	Plastic materials and synthetic resins	282
6	Agricultural chemicals	287
7	Soap, detergents, cleaners, perfumes, cosmetics and toiletries	284
8	Paint, varnishes, lacquers, enamels and allied products	285
9	Miscellaneous chemical products	289

NUMBER	PRODUCT FIELD	SIC CODE
10	Drugs and medicines	283
11	Petroleum and natural gas extraction and refining	13, 29
12	Rubber and miscellaneous plastic product	30
13	Stone, clay, glass and concrete products	32
14	Primary ferrous products	331, 332, 3399, 3462
15	Primary and secondary non-ferrous metals	333–6, 339 (except 3399), 3463
16	Fabricated metal products	34 (except 3462, 3463, 348)
17	Engines and turbines	351
18	Farm and garden machinery and equipment	352
19	Construction, mining and material handling machinery and equipment	353
20	Metal working machinery and equipment	354
21	Office computing and accounting machines	357
22	Special industry machinery, except metal working machinery	355
23	General industrial machinery and equipment (included refrigeration and service industry machinery)	356, 358
24	Electrical transmission and distribution equipment	361, 3825
25	Electrical industrial apparatus	362
26	Household appliances	363
27	Electrical lighting and wiring equipment	364
28	Miscellaneous electrical machinery, equipment and supplies	369
29	Radio and television receiving equipment except communication types	365
30	Electronic components and accessories and communication equipment	366–367
31	Motor vehicles and motor vehicle equipment	371
32	Ship and boat building and repairing	373
33	Railroad equipment	374
34	Motorcycles, bicycles and parts	375
35	Miscellaneous transportation equipment	379 (except 3795)

NUMBER	PRODUCT FIELD	SIC CODE
36	Aircraft and parts	372
37	Professional and scientific instruments	38 (except 3825)
38	Other manufacturing	All other SIC's

Notes

1 For a thorough analysis of the theoretical and empirical aspects of intra-industry trade, see the by now classic work by Greenaway and Milner (1986).
2 The latter has been confirmed by a recent study (Gerstenberger, 1990) focussing on the evolution of trade specialisation on the basis of the largest commodity disaggregation made possible by international trade data (the five digit of the Standard International Trade Classification). It shows that the similarity in the export vectors of the United States, Japan, Germany, France and Great Britain decreased (considerably in some cases) between 1978 and 1986. The exceptions are a greater convergence between Japan and Germany, on the one side, and between Japan and the United States, on the other.
3 As documented by Cantwell (1989), there have been significant changes in the international patterns of technological specialisation over the last century. These changes have been, however, affected to some (limited) extent by the initial structure of technological activities.
4 The structure of technological advantages in Soete's work is based on patenting in the United States from 1963 to 1977; the structure of trade advantages is based on exports in 1977.
5 For this purpose, the US Patent and Trademark Office (PTO) has established a concordance. The concordance assigns patent sub-classes to all the product fields to which they are pertinent. In order to avoid multiple patent counts among product field categories, we referred to the fractional patent data set established by the PTO.
6 Formally the RCA indicator is given by the following equation:

$$RCA = \frac{X_{ij}}{\Sigma_i X_{ij}} \Bigg/ \frac{\Sigma_j X_{ij}}{\Sigma_i \Sigma_j X_{ij}}$$

where:
X_{ij}=exports of product j of country i;
$\Sigma_i X_{ij}$=world exports of product j;
$\Sigma_j X_{ij}$=total exports of country i;
$\Sigma_i \Sigma_j X_{ij}$=total world exports.
7 Formally, the CTB indicator of a country i relative to a product xxw is given as:

$$CTB = \frac{X_j - M_j}{(X+M)/2} \times 100 - \frac{X-M}{(X+M)/2} \times \frac{X_j + M_j}{(X+M)} \times 100$$

where:

X_j=exports of country i of product j
M_j=imports of country i of product j
X=manufacturing exports of country i
M=manufacturing imports of country i

8 The asymmetry between the technological and the production diversification of firms is a result of the fact that only a part of upstream (in capital goods), horizontal (in mechanical and electronic instrumentation) and downstream (in consumer goods) technological activities are transformed into corresponding productive activities. This phenomenon is largely explained by the need for each firm to develop technological interdependencies with its suppliers, partners and clients (Pavitt et al., 1989). In other words a firm's technological frontier extends beyond its productive frontier.

9 The same interpretation has been suggested by Cantwell (1989).

10 A higher concentration of technological specialisation in smaller countries has also been found by Soete (1987) and Archibugi and Pianta (1990). Amendola and Perrucci (1990) also obtained analogous results with respect to the trade specialisation in high-tech products of European Community countries.

11 As regards technological specialisation, the lowest values are found for the United States. However, since a large share (approximately 66 per cent on average) of US patents in each sector are granted to US firms or individuals, the result is that the American specialisation profile is 'flat,' with only minor differences between weak and strong sectors. Therefore, the degree of concentration of US technological specialisation given in table 7.1 must be considered as purely suggestive.

12 Comparisons were carried out between the first and second period, between the second and third period, and between the first and third period.

13 That does not necessarily imply a reduction in the country's degree of specialisation. It can, in fact, be shown that $\sigma_t^2/\sigma_{t-1}^2 = b^2/R^2$, where σ^2 indicates the variance, b the regression coefficient and R^2 the Pearson correlation coefficient. For values of b between 0 and 1, b can be, and often is, greater than R. In such a case, the degree of specialisation of a country increases. This situation arises when the decrease in the degree of specialisation induced by the 'regression effect' ($0<b<1$) is more than offset by the 'mobility effect' (Cantwell, 1989) of sectors within the existing ordering. The 'mobility effect' implies important changes in the initial ranking of sectors. For more details on the methodological aspects of the testing procedure adopted here, see Cantwell (1989 , pp. 25–31).

14 In these sectors, French firms often rank among the top ten in the world in terms of patenting capacity. They include numerous state-owned firms and state agencies that have prospered in a 'Colbertist' tradition (Pavitt and Patel, 1990).

15 Parameter b is also significant at the 5 and 1 per cent in, respectively, 57 and 31 per cent of the equations estimated.

8 Trade on high-technology markets and patent statistics – leading-edge versus high-level technology

HARIOLF GRUPP AND GUNNAR MÜNT

1 What high technology is understood to mean

International technological competitiveness seems to become an ever more important issue (Porter, 1990). Between the American and the Japanese economic superpowers the countries of the European Communities are broadly diversified with old traditions in research and development (R&D). But exactly what is the triad countries' strength in high technology trade? There is no straightforward answer to this question as 'high technology' is not a well-defined issue in economics. It is to a significant extent a public and political manifestation, and, therefore, not capable of easy or precise definition in terms of economics or innovation research (Johnston and Hartley 1986).

By some authors high technology is defined by its R&D intensity. Thereby, R&D expenditures divided by turnover are taken as the benchmark for classification. However, this definition of R&D-intensive items is problematic as well. Most common is still a classification of industries according to their R&D expenditures per turnover and a bipartition of all products or services these industries are manufacturing or providing, into high technology or not. (This definition is preferred by the OECD, the US Department of Commerce and other public bodies.) Yet, the present techno-economic system is characterised by the increasing significance of technology fusion or spillover effects (Kodama, 1991; Grupp, 1991, 1996; see also other references cited by these authors). Industries cannot represent homogeneous technologies. A way out of these difficulties is an analysis of markets, not industries, and therefore the consequent use of product-based R&D intensities (Abbott et al., 1989; Legler, 1987; Grupp, 1991). But this alternative procedure is still heavily debated (United States International Trade Commission, 1990).

Since the International Trade Classification has undergone a complete revision, from 1988 onwards the more traditional approaches to define high technology may not be continued on an international level. Therefore, the Fraunhofer Institute for Systems and Innovation Research (FhG-ISI) designed a new list of R&D-intensive products (Legler *et al.*, 1992). This list is modelled after an earlier approach by the Lower Saxony Institute for Economic Research (NIW) which was based on the previous International Trade Classification (Legler, 1987).

Table 8.1 lists the product groups considered to fall into the category of high technology. The products in table 8.1 are defined by the Standard International Trade Classification (SITC), Revision III. Starting from the micro-level of individual firms, their R&D expenditures and their turnover in major OECD countries at the end of the eighties were sorted along with their R&D intensity. Because of lack of internationally comparable data, national sets of data were used instead. Company R&D statistics are available for the United States (Standard and Poor's Compustat Services) and for Germany (Schwitalla, 1993) and product-based national statistics for Japan, the US, Germany, Italy, Sweden and the Netherlands (see Grupp, 1991). The following cut-off rules were derived from frequency analysis: all products with R&D intensities above the industry average of about 3.5 per cent of turnover are included. This is understood to be *high technology*.

A particular advantage is the fact that this list divides the R&D-intensive sector into two parts, differentiating between *leading-edge products* and products forming part of *high-level technology*. The threshold between the two categories is also derived from a frequency analysis of R&D intensities from the above-mentioned national data and is allocated at 8.5 per cent of turnover.

Recently, a similar product-based list of high technology has been promoted by Amendola and Perrucci (1990). By stressing the limitations of sectoral R&D intensity classifications, these authors based their selection of products on the NIMEXE nomenclature used by the European Community. This nomenclature is more disaggregated than the SITC III classification used in this chapter. Specific inquiries were conducted in manufacturing firms by asking experts to evaluate products on the basis of a number of parameters including R&D intensity, the degree of automation in production, characteristics of product use and product life cycle. In this way, 254 high technology products on a six-digit level of disaggregation were defined (Amendola and Perrucci, 1990). By comparing these products with the definition represented by table 8.1, one finds that more than 70 per cent of them fall into the category of leading-edge technology and only less than 30 per cent into high-level commodities. Thus, it is concluded that this newly designed bottom-up definition of high technology may be equated more or less with *very* R&D-intensive production.

Table 8.1. *R&D-intensive products and government intervention*

h:	SITC III	Product group (non-official terms)	PP	GP	GS	BE	PR	ER
Leading-edge technology								
1	516	Advanced organic chemicals	■					■
2	525	Radioactive materials	■	■		■		■
3	541	Pharmaceutical products		■		■	■	■
4	575	Advanced plastics						■
5	591	Agricultural chemicals				■	■	■
6	714	Turbines and reaction engines	■	■				
7	718	Nuclear, water, wind power generators	■	■	■	■	■	■
8	752	Automatic data processing machines	■	■				
9	764	Telecommunications equipment	■				■	■
10	774	Medical electronics	■					
11	776	Semi-conductor devices	■	■				
12	778	Advanced electrical machinery	■					
13	792	Aircraft and spacecraft	■	■				
14	871	Advanced optical instruments	■	■				
15	874	Advanced measuring instruments	■	■				
16	891	Arms and ammunition	■	■				
High-level technology								
17	266	Synthetic fibres						
18	277	Advanced industrial abrasives						
19	515	Heterocyclic chemistry					■	
20	522	Rare inorganic chemicals						
21	524	Other precious chemicals					■	
22	531	Synthetic colouring matter					■	
23	533	Pigments, paints, varnishes						
24	542	Medicaments	■			■		
25	551	Essential oils, perfume, flavour					■	
26	574	Polyethers and resins					■	
27	598	Advanced chemical products					■	
28	663	Mineral manufacturers, fine ceramics						
29	689	Precious non-ferrous base metals	■					
30	724	Textile and leather machinery						
31	725	Paper and pulp machinery						
32	726	Printing and bookbinding machinery	■					
33	727	Industrial food-processing machines					■	
34	728	Advanced machine-tools						
35	731	Machine tools working by removing						
36	733	Machine tools without removing						
37	735	Parts for machine-tools						
38	737	Advanced metalworking equipment						
39	741	Industrial heating and cooling goods					■	

Table 8.1 (*cont.*)

h: SITC III	Product group (non-official terms)	PP	GP	GS	BE	PR	ER
High-level technology							
40 744	Mechanical handling equipment						
41 745	Other non-electrical machinery						
42 746	Ball and roller bearings						
43 751	Office machines, word-processing				■		
44 759	Advanced parts for computers				■	■	
45 761	Television and video equipment					■	
46 762	Radio-broadcast, radiotelephony goods					■	
47 763	Sound and video recorders					■	
48 772	Traditional electronics						
49 773	Optical fibre and other cables				■		
50 781	Motor vehicles for persons					■	■
51 782	Motor vehicles for goods transport						■
52 791	Railway vehicles	■				■	■
53 872	Medical instruments and appliances				■		
54 873	Traditional measuring equipment						
55 881	Photographic apparatus and equipment						
56 882	Photo- and cinematographic supplies						
57 884	Optical fibres, contact, other lenses						

Notes:
PP: Large share of public production.
GP: Considerable government procurement.
GS: Noticeable government subventions.
BE: Barriers to market entry or production quantity.
PR: Strong price regulation.
ER: Environment regulation.

Why such a distinction between very R&D-intensive products (leading-edge technology) and (above-average) high technology (high-level commodities)? The definition given may be viewed in a converse fashion: a very high percentage of expenditure on R&D signifies low turnover expectations. Indeed, for every million spent on R&D in leading-edge technology, the average turnover is less than 12 million, while a typical figure for high-level technology is 30 million and more for each million invested in R&D (Grupp, 1991). But again, why is the turnover expectation for leading-edge technology lower than for the high-technology category? Table 8.1 also includes some rather impressionistic classifications by six modes of government intervention.

For some product groups a large share of public production (PP) is observed, whereas other product groups may be characterised by strong government procurement (GP). It is undeniable that energy generation, aircraft and spacecraft and other products are subject to government subsidies (GS) in many countries. For other product groups, barriers to market entry (BE) or production or at least export quantity are the case, which is certainly true for the pharmaceutical sector requiring approval procedures by the respective food and drug administrations or health care authorities. Consumer electronics along with cars, steel and textiles have seen at various times the application of Voluntary Export Restraints, technical barriers to trade (e.g., colour television systems) and the like (Cawson and Holmes 1991, p. 1973). Strong price regulation (PR) is at least the case for railway and telecommunications services whereas environmental regulation (ER), as a more recent form of government intervention, is becoming more and more pervasive throughout the economy.

It is not intended in this chapter to discuss the justification for government intervention, rather the point is made that the level of government intervention affects leading-edge technology more than high-level technology (see table 8.1). Leading-edge technology includes many products that are subject to protectionism, such as aeronautics and aerospace or nuclear energy, whereas most of the high-level commodities are mass consumption products traded internationally more or less under free-market conditions. Thus, the distinction between the two high-technology categories seems to be meaningful at least in terms of a cognitive model for further studies of the various functions of national science and technology production and its impact on competitive trade advantage. Analogous deliberations with respect to high technology, procurement, trade barriers and regulation in the Single European Market after 1992 may also be found elsewhere (Buigues *et al.*, 1990).

2 Competitiveness, technological asymmetries and absolute advantages

International competitiveness is an elusive concept. It is not the same as the microeconomic notion of competition on markets. As Krugman (1991b, p. 811) puts it, 'a country is not much like a business. Indeed, trade between countries is so much unlike competition between businesses that many economists regard the word 'competitiveness', when applied to countries, as so misleading as to be essentially meaningless. . . . International competition does not put countries out of business'.

Thanks to the more recent valuable insights of evolutionary, new growth and trade models, however, it is commonly understood that there is a close relationship between a nation's ability to generate and sell technology-intensive products at home and abroad and its overall growth record. In

general, the realisation of the high-growth potential of goods is supposed
to reflect a nation's competitiveness. From its early beginnings, evolutionary economics has focused on the innovation-growth linkage. This school
of economic thought is essentially based on the work of Schumpeter (1911,
p. 99). The extension of Schumpeterian concepts to trade theory was in
particular laid down by Posner (1961) and Lorenz (1967) among others.
According to this thinking, a country rich in innovation will create a technological lead, or 'technology gap', compared with those competing countries less active or less successful in producing innovations. The gap
determines to a large extent the structure of traded goods as well as
national production. Obviously, if the aforementioned argument holds
true, technological leads are linked with domestic growth and national
trade structure through various feedback mechanisms – and thus influence
what might be termed competitiveness. But what is the exact mechanism
underlying this process?

This contribution neither pretends to develop a comprehensive theoretical tool to tackle this web of complex relationships, nor to deliver an
empirical test of different trade models on the basis of available trade and
technological indicators. Instead, we intend to outline the more basic theoretical and empirical underpinnings of what is normally understood to be
the concept of (technological) competitiveness. As there is neither a clear-
cut normative measure nor an analytical tool for easily handling the
concept of competitiveness, one usually goes back to the second-best (normative and analytical) concept available. Thus, the focus of interest is normally directed to the relationship between the *generation of high-technology
products, international trade and economic growth.*

Whereas economic growth and subsequent welfare gains are illustrated
in detail by various economic indicators, the impact of high-technology
production on international trade cannot yet be accounted for analogously
by innovation and trade indicators. The measurement of international
technological activity is almost entirely reduced to patent statistics. With
the contributions of Pavitt and Soete (1980), Soete (1981, 1987), Legler
(1987), Dosi *et al.* (1990), Legler *et al.* (1992) and Grupp *et al.* (1992) there
is, however, a lot of evidence at hand that indicates that a strong relationship between an economy's technological and trade *specialisation* exists.

As far as the effects of *technological specialisation* on economic growth
– and, thus, a nation's competitiveness – are concerned, evidence is scarce,
so far.[1] Although we know about the overall positive relationship between
the levels of innovativeness and economic activity as shown in the before-
mentioned contributions of Fagerberg (1988a), Dosi *et al.*, Legler *et al.*
(1992) and in the recent article by Amendola *et al.* (1993), we lack an exact
understanding and evidence about the relationship between a nation's

distribution of technological activity on selected technology fields and the induced effects on international trade and economic activity in the corresponding sectors. That means, our knowledge about what lures R&D funds away from some products into other technological fields, what brings about innovation in the latter fields and what are the implications this has on growth and competitiveness is, at best, incomplete.

This is why we have to rely, up to now, on theoretical models describing the interaction between technological and economic change. Traditional (neo-classical) economic theory conceives international trade as an all-over adjustment process which links the world-wide national endowments with production factors via the price–quantity relationship. In neo-classical theory the focus is on the efficiency of the static allocation of resources; in trade theory this notion is extended to the question of whether there are welfare gains from trade in opening up economies to the international division of labour. Thus, there is only an intermediate link between the (static) specialisation due to comparative advantages and the dynamic process of economic growth. In its most basic version, comparative differences between countries are either the result of differences in consumer preferences, production technology or factor abundance. Differences in utility functions are normally assumed to vanish; country-specific production functions are usually difficult to handle in empirical verifications of the hypothesis. Consequently, under the assumption of internationally identical production and utility functions, neo-classical trade theory has often been reduced to the elaboration of the Heckscher–Ohlin hypothesis which explains the observed trade structure according to comparative advantages from the country-specific abundance of production factors.

Whether useful for the analysis of modern trade-related issues or not, this factor-abundance hypothesis raises a lot of obstacles when looking at the mechanisms that govern the structure and development of international trade in high technology goods. Together with the accompanying assumptions about profit-maximizing behaviour and identical preferences, the analysis of structural change in neo-classical trade theory can be reduced to the adjustment mechanism of relative prices and quantities. Thus, processes of international specialisation in technology and trade have to be squeezed into the framework of factor-abundance analysis. To what extent international or intersectoral differences in technology can be dealt with on the basis of this theoretical background cannot be discussed here. Suffice to mention that a large body of research on the economics of innovation has revealed the rather complex web of relations – independent of the underlying resource endowment – that govern the development of technology and its impact on trade structure.[2] Consequently, reducing the analysis to changes in relative prices and their adjustment mechanisms according to internationally different resource endowments might strip off from an analysis the most interesting insights into the determinants of technological competitiveness of nations.

One of the first, and until now, most outstanding intellectual contributions to the problems of the impact of technical change on economic growth and trade structure has been offered by Kaldor (1957, 1962 (with Mirrlees), 1978, 1981) which have been extended to open economies by Thirlwall (1979). Based on neo-Keynesian ideas, Kaldor stressed the mutual feedbacks between the rates of economic growth and productivity that give rise to dynamic processes of 'cumulative causation' and virtuous circles of single economies or sectors. Thus, once gained, productivity advantages will be followed by long-term rises in the rates of growth – which accelerate productivity growth in turn. Ultimately, these macro-economic dynamics will lead to structural changes in favour of those industries caught in virtuous circles. International trade, on the other hand, might serve – in this case through high demand potentials – as an accelerator in this process.

Dosi and Soete (1988) have taken up Kaldor's ideas and combined them with Schumpeter's theory of economic growth stirred by continuous innovation. Based on a broad empirical body of literature, they set out a conception that accounts for persistent differences in the rates of innovation in single technological fields and the corresponding economic sectors. The general heterogeneity in the dynamics in the rates of technical change is assumed to be reflected in different paths of macroeconomic or sectoral growth. Consequently, differences in economic structure lead to differences in national technological and economic activity. Then, sectoral and trade specialisation follow the patterns of growth rate differentials in technical and economic change which are themselves induced by the processes of cumulative causation at a sectoral or product level. In order to assess the implications of diverging rates of technological activity on trade and growth in single economies, we have to look at the structure and persistence of technology gaps between countries.

If innovation and growth potentials persistently differ between technological fields and sectors, countries could evade being dragged into vicious circles of constant decline through structural change of the national economy. Evidence, however, militates against this view because a country's economic structure – built up over a long time, resulting in various different institutions and influenced by national policies – limits the possible rate of economic change to a more sluggish modification of the existing network between firms, public research institutions and government. The basic drive for economic change is, thus, the country-specific adoption of different technological and economic paradigms which are adapted to the already existing national innovation structure. By comparison, radical structural change which opens up totally new opportunities is a rather rare phenomenon. Consequently, technical change is neither a free good nor is it costlessly transferred between different countries but is shaped and generated in different country-specific ways. Especially, it is the cumulative process of learning that gives a country the possibility to acquire technological knowledge that is distinct from that of other nations. Following the accumulation of this specific knowledge in single technological fields, an economy is able to allocate R&D funds to areas of built-up national advantages, more or less independent of any endowment with natural resources. The structural approach applied here – whereby structure is defined as distinct from the traditional sectors either as products or technological fields – allows for different technological potentials built up in a process of continuous technological learning in innovation and production.[3]

Returning to the concept of technological competitiveness, *structural competitiveness* is quite apart from traditional neo-classical notions. Following Dosi and Soete (1988), we regard comparative national advantages as being dominated by *absolute advantages*. Although similar to the mathematical term of neo-classical comparative advantages, they are quite distinct from this in meaning.[4] Absolute advantages act as the main determinant of international trade in the case of different rates of technical change between countries. As mentioned earlier, in neo-classical theory comparative advantages are mostly based on differences in relative (autarky) prices that will be equalised in the course of international adjustment and instant restructuring of the economy. In contrast to this, absolute advantages result in different rates of macroeconomic or sectoral activity which is due to long-term technological advantages caused by processes of cumulative causation. Thus, instead of looking at international equilibria of factor prices one has to focus on persisting differences in the rates of technological and economic development.

Although we know that specialisation in international trade is made up of a variety of different determinants such as exchange rates, national policies or prices for commodities, the influence of differences in technical change between countries prevails throughout.[5] If, however, technological asymmetries persist between countries, long-term adjustments of macroeconomic activity and not changes in relative prices will be the result. According to traditional trade theory, price adjustments in favour of the products of the technological leader will bridge the productivity gap. This rather optimistic view of international economic development continuously heading towards a general equilibrium is refuted by a simple model taken again from Dosi and Soete (1988): Suppose the technologically lagging country never succeeds in catching-up through her own innovative efforts. Unless there is a massive transfer of technology to the lagging country, the technology gap is never closed. In the long run, price adjustments, structural change and specialisation in the products of comparative national advantage will not restore the country's competitiveness. Consequently, world demand for the lagging country's products vanishes and macroeconomic activity, especially employment, falls in turn.

Relaxing these somewhat crude assumptions on overall technological leads of one country across all products, we are especially interested in the question for which technological specialisation profile a country should strive in order to establish these absolute advantages due to technological leads – and thus reaching higher levels of macroeconomic activity. It goes without saying that countries, according to their absolute size, cannot engage themselves in all technological fields with the same amount of resources spent on R&D and production. Thus, economies have to specialise economically and technologically – that is, select some areas and skip others. Which technological or product fields a country chooses for specialisation depends on a variety of different factors. 'Natural' abundance of production factors, accumulated technological knowledge and skills in production, and the realised specialisation of competitors on world markets rank among the most important.

Returning to the focus of this chapter, we want to verify the relationship between the technological activity of countries in some selected high-technology fields and their corresponding world market share which fit best the concept of demand-driven virtuous circles between productivity dynamics and growth – and, thus, absolute advantages. There are several indicators available. Largely consistent with both theoretical concepts of competitive or absolute advantages, we consider world export shares to be the best-fitting indicator. It is defined as follows:

$$WES_{ikt} = (ex_{ikt} / \Sigma_k ex_{ikt})$$

Therein ex_{ikt} denotes the exports of country i in product group k at time t. The WES indicator measures what level of economic activity a country has reached on the world market for a special product. A larger share of world exports hints at a higher level of competitive or absolute advantage compared with other countries. Following our above argument, we suppose that higher levels of advantage in certain world market segments (product groups) correlate with higher levels of technological activity as evidenced by output of patents. Thus, we take the *absolute number* of a carefully selected set of patents (for a definition of sets see section 4) of a country in single technological fields, and relate them to all triad patents in this field. The world patent share is denoted by WPS_{ikt} with the above subscripts.

Further we want to find out how *technological productivity* in high technology areas, as defined in section 1, affects (technological) competitiveness and relates to the world market shares in the (more regulated) leading-edge and the (less regulated) high-level areas. To this end we adjust the patent indicator according to the size of a country's labour force. This indicator is defined as follows:

$$PL_{ikt}=(P_{ikt}/L_{it})$$

PL_{ikt} denotes the patents per labour force (L_{it}) in country i and technological field k at time t. Which sets of patents should be preferred for reasons of adjusting for different regional patent propensities, will be described in detail in section 4. Adjusting for the size of the labour force leads to patent intensities (or 'technological productivities' per worker). What we are interested in, is the question whether single countries – by specialising in specific high-technology fields – can increase their world market shares in high-technology areas correspondingly.

Thus, if the model of (absolute) technological competitiveness applies to all high technology product categories (or technological areas) the same way, we will expect a positive correlation between world market shares and technology intensity. Before presenting our results we stay with the related problem of national control of technology in the following section.

3 National control on R&D or globalisation?

In face of techno-globalism national technology and industrial policies are under pressure. In the future more so than in the past, national policies will have to compete both with supra-national bodies (like the Commission of the EU or the NAFTA bodies) and with regional activities (state and province policies like in the US, Germany, Italy or Spain and local activities, e.g.: campus-based incubator centres). Yet, on the firm side, multinational direct investments, global sourcing, networking and the 'mobile' location of R&D establishments has progressed more rapidly than the related government policies.

This 'globalisation' would suggest including all countries of the world in such an analysis. However, theoretically, it is more correct to consider only bilateral trade relations. After all, it is by no means certain – indeed there is some evidence to the contrary – that the advanced economies always supply products of the same high technological content. That is why a consideration of trade relations within groups of similar economies is preferable (Grupp, 1991, p. 281). As a compromise, we restrict our analysis to the 18 larger OECD economies (see figures 8.1 and 8.2 in section 5), other than Daniels (1993) who included 52 countries in his data.

With this selection of countries, the empirical task is to measure their disembodied knowledge by patents. By bringing together patent data and trade data, one main reservation must be discussed. It relates to the extent to which the variables reflect R&D processes under national control (Grupp *et al.*, 1992). Problematic is in particular the strategic control of technology generated by patents (Grupp *et al.*, 1992). In most cases, inventors do not decide when, to what extent, for which markets, and for which products their inventions are used. Nor do they have a say, in general, on licensing, selling or abandoning of patents, insofar as they are employees of commercial firms. Consider, for instance, Belgium: Belgian technology is strongly controlled by non-Belgian companies. Patel and Pavitt (1991a) reported that nearly 40 per cent of Belgian patenting in the United States is due to large foreign-controlled firms, whereas for Western Europe a comparable figure is 6 per cent, on average. Belgium had the largest share of national patents generated by foreign capital in their sample of 11 countries.

Even more problematic is the foreign trade indicator WES. It is by no means certain that high-technology exports are built on technology and science indigenous to the exporting country. The public good 'R&D results' is highly mobile. In particular, large multinational corporations may produce and export from countries other than those where the scientific and technological achievements originate.[6] The degree of high-tech trade advantage will thus depend not only on the existence of scientific or technology gaps, but also on the number and nature of high-tech firms that manufacture in that country, and thus on the degrees of opportunity, cumulativeness and appropriability (i.e., interfirm technological capabilities) of the local export sector (van Hulst *et al.*, 1991).

As we study major OECD countries, we have to include the countries of the European Union (EU). This creates another problem of globalisation: One observes increasing amounts of intra-EU trade (see Grupp and Soete, 1993, pp. 58–64). At least in the information industry, trade diversion seems to have taken place alongside the more recent phases of economic integration within the single market. If we consider this as regional trade, then the real international share for some countries will cover less than 50 per cent of their respective exports. For other OECD countries, like the USA or Japan, the regional and overseas share in exports are quite different.

These reservations should be kept in mind when both patent data and foreign trade data are put into the perspective of national assignments.

4 A simple triad model for OECD countries

Private or corporate research generally produces patents rather than academic publications (Dasgupta and David, 1987; and Grupp, 1990). Patent statistics are an accepted output indicator for codified knowledge from strategic and applied research and industrial development. As patent applications are legal documents that are valid in only one country, many foreign 'duplications' of domestic priority patent applications are generated. The selection of patent data from only one patent office, therefore, does not always yield an indicator that is representative of the world output of inventions. As patents can be matched to trade categories if the patent classification of the United States is being used (Legler *et al.*, 1992), the classification as disclosed in patents issued by the United States Patent and Trademark Office are examined further for a breakdown of technology.

Annual averages of granted patents for the invention years 1986 through 1988 have been selected. These patents may be regarded as proxies for corporate attempts to protect their goods produced in 1990. Among others, Legler *et al.* (1992) and Amendola and Perrucci (1992) found in different types of investigations, that patent statistics, because of the cumulative nature of innovation, precede international trade by about three years. Accordingly, the trade data from 1990 are selected.

The patents granted in the United States, from which the product classification is taken, may serve as a benchmark for an analysis of the bilateral trade between any one OECD country and the United States. It is questionable, however, whether *international* trade (also within the EU and between the EU and Japan) should be reflected in foreign intellectual property rights in the United States. From a theoretical perspective this is an insufficient condition. Therefore, another demarcation of intellectual property rights with significance for international trade was used.

As duplications of patents can be traced and matched to each other, so-called patent families may be defined centering around one invention and bringing together the foreign property rights in all countries of the world. The selection criteria in the investigation which follows was that only those inventions are taken into consideration where a foreign duplication at least in the United States, in Japan and at the European Patent Office was filed. By this selection criteria, the *'triad' model* is applied requiring protection of industrial property in each of the triad blocs USA, Japan and Europe. As the condition requires that in the United States a patent is not only filed but also granted, the classification may be taken from the US equivalent and matched to the product group of potential application. Thus, the patents were assigned to either high-level or leading-edge products by a respective patent-to-sales concordance (Legler *et al.*, 1992). This is considered as a major achievement to cover the world's output in technology instead of using solely US patent data (the standard procedure; recently see Frame, 1991, and Dosi *et al.*, 1990). Yet, an elaboration of this method is not possible within this chapter (see Grupp, 1994a).

In order to test the usefulness of the triad concept, we examine the world market share for all manufactured products in relation to various sets of patent data, among them the triad patents. Further, we assigned the US patents by inventors' countries (residence of inventors) as well as by assignees' countries (first application principle) to get an impression of the globalisation effects. The first set (USINV) represents the country of origin, the second the firm/country controlling the technology (USASS). Next, we selected Japanese patent applications (at the Japanese office Tokkyo cho – TC), patent applications at the German Patent Office (DPA) and patent applications at the European Patent Office (EPO).

From these data sets, we constructed three more patent samples (all sets being adjusted to 1986–8 priority years also for granted US patents; we did not use patent dates). We merged all domestic patent applications for the OECD countries (DOM), we added the EPO patents with destination Germany to the DPA data (and denote this set as GE for Germany/Europe), as legal protection on the German territory is warranted on both ways, and finally selected the triad patents TP as described above.

5 Discussion of empirical results

By linear OLS regression of the data sets (in each case the world patent share per set) with the world export share WES in manufactured goods we found the results as compiled in table 8.2.

Two sets of patents are highly significantly correlated with the export performance, the triad patents (as expected), but also the European data. Both tests are heteroscedasticity robust and also hold in Spearman rank correlations. The European Patent Office requires fees which are considerably higher than at European national patent offices so that cost-effectiveness is only achieved if the application is designated to three or more countries under the European Patent Convention. From this, there seems to be a similar selection process of the most important patents with relevance for world markets as by the triad model.

The US patent data explain the trade advantages reasonably well (significant correlation) and there is nearly no difference between the inventors' countries and the assignees' countries which downplays a strong distortion effect from globalisation. However, the Durbin–Watson statistics are inconclusive; an autocorrelation of the data cannot be rejected. Reasonably good is the correlation when all patents with protection in Germany are taken (GE), whereas for the patent documents at the German Patent Office alone (DPA), neglecting the European access to protection, the same correlation gets weak. Domestic patents are not a good reflection of world trade nor do the patents in Japan.

Table 8.2. *Regression results for world export share (1990) and seven patent samples (priority years 1986–90)*

Patent sample	R^2 (adj.)	F	t	a/%	DW	n
USINV	0.41	12.0	3.46	0.35*	1.07	17
USASS	0.39	11.3	3.36	0.43*	1.04	17
TC	0.09	2.6	1.62	0.13	2.76	17
DPA	0.38	6.6	2.57	3.32*	2.12	10
EPO	0.87	108.3	10.41	0.00***	1.48	17
DOM	0.22	5.6	2.38	3.13*	2.59	17
GE	0.60	14.4	3.79	0.53**	2.48	10
TP	0.59	24.0	4.90	0.02***	1.33	17

Notes:
*** highly significant at the 0.1 per cent level; ** significant at the 1 per cent level; * weakly significant at the 5 per cent level.

From this we conclude that higher levels on certain world market segments, in manufactured goods, fall together with higher levels of technological activity as evidenced by patent output. But due to the limitation of patent property to national markets, the analysis of high technology trade advantage should be based either on European patents or on triad patents. As a concordance of the International Patent Classification, applied by the EPO, to the SITC is not available we chose the triad patents and the match to trade categories by the US product classification.

Linear regression calculation was done in order to investigate how technological productivity relates to world market shares in high technology

$$WES_{ik(1990)} = b + m \cdot PL_{ik(1986-88)}$$

For k we selected

all high technology products and related patents as defined in table 8.1,
all leading-edge products and related patents as defined in table 8.1,
all high-level products and related patents as defined in table 8.1.

Figures 8.1 and 8.2 provide an overview over the technological productivity of OECD countries.[7] The correlation of those international 'triad' patents with potential applications in high-level commodities matched to the respective trade data is weaker than in table 8.2 for all technologies (OLS regression with $R^2 = 0.24$ and $t = 2.45$ is significant at the 5 per cent level with $a = 2.71$ per cent; the Durbin–Watson test is inconclusive

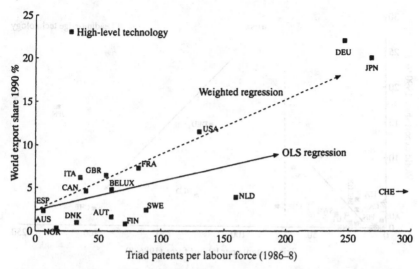

Figure 8.1 Relation of international 'triad' patents (per labour force 1986–8) and world export share (1990) distinctly matched for high-level technology
Notes: Country acronyms according to the three digit ISO convention – DEU: Western Germany; DNK: Denmark; FRA: France; ITA: Italy; GBR: United Kingdom; NLD: Netherlands; ESP: Spain; IRL: Ireland; GRC: Greece; BELUX: Belgium and Luxemburg; PRT: Portugal; DDR: Eastern Germany, former German Democratic Republic.

with DW=1.19). The relation is heteroscedasticity robust; if we take the inverse labour force as a measure of variance, i.e., we assume the data of the larger countries to be more precise, then the weighted regression is excellent (R^2 (adj.)=0.72, t=6.54, DW=1.74, α<0.1 per cent). The most notable disparity is observed for Switzerland, a very productive (in terms of technology), but isolated country in the geographical centre of Europe's economy, a country which does not want to join the European Union, but is surrounded by member states. In the weighted regression, the weight of the data of this and other smaller countries with larger variances is reduced for a variety of reasons (see section 3). These countries cannot reach the expected world market shares corresponding to their technological productivity (for an optical impression see figure 8.1).

Yet, there is no significant correlation for those remaining patents with codified knowledge for potential use for leading-edge technology (linear regression with R^2=0.18 is not significant at the 5 per cent level; the relation is positively autocorrelated with DW=0.95 and not heteroscedasticity robust). If we bring both high technology sectors together the correlation holds again on a weak level (R^2=0.25; DW=0.96 points again to autocor-

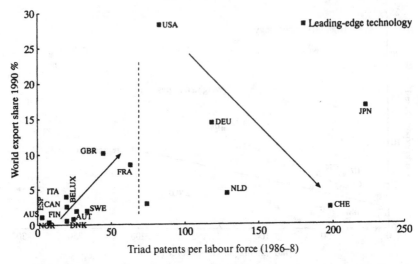

Figure 8.2 Relation of international 'triad' patents (per labour force 1986–8) and
world export share (1990) distinctly matched for leading-edge technology
Notes: For country codes see figure 8.1

relation, which is 'imported' from the leading-edge markets into the full
sample of high-technology products, no figure).

Technology production expressed by intellectual property rights seems
to have a very important function for structural advantages in the more
market-oriented parts of high technology. Despite the reservations made in
section 3 with respect to the control of technology by patents which may
easily cross borders, overall the national factors are strong enough to maintain a strong coupling between the local source of the invention and the
national share in world trade, in particular for the larger countries. There,
technology is presumably controlled by national companies.

For leading-edge technology the situation is more differentiated. Only to
guide the eye, we inserted two lines in figure 8.2. They do not represent
regression lines. With too little patents a nation has no chance of competing on the international markets for technology. If technology output is
more productive, like in France and the UK, absolute advantages on international markets accrue. But above a certain threshold, the protected
leading-edge markets are controlled by other means than intellectual property rights. Leading-edge technology is characterised by many distorting
factors from protectionism, regulation and government intervention. These
factors are building up in such a way that the more direct function between
technology produced and trade success achieved is vanishing. The world

leader in leading-edge exports are the United States on a medium-technology production level. Germany, the Netherlands, Switzerland and Japan try hard to catch up in leading-edge technology – but with little success in trade so far.

6 Conclusions and an outlook on national specialisation

As a result of this investigation, different structures between technology production and competitive trade are laid open. National systems of innovation continue to be very strong which make simplistic investigations difficult. Globalisation of modern technology, as important as it may be, does not relieve national technology policy of responsibility for local R&D activities and innovative industrial sites.

One has to bear in mind that this chapter contains a first exploration of international trade analysis based on other disaggregation procedures than by industries. The exercise to split the high-technology markets into two parts seems to be very helpful for an in-depth analysis of national strengths and weaknesses. The rationale behind the two-piece list of R&D-intensive markets used in this chapter is that there are two hemispheres in the world of R&D intensity. One (the high-level markets with expectations of a relatively good turnover per R&D investment), in which technological production originating from the R&D activities by local industry, plays a role and is a decisive factor for international competitiveness. Scientific achievements are not so important here. The other hemisphere (the leading-edge markets with moderate expectations in turnover per R&D investment), in which factors other than technology guarantee international success, is characterised by stronger government intervention both on the side of R&D and also in terms of procurement and regulation. Here, national technology production is not fully appropriated by local enterprises and not converted into world market shares.

It is a severe limitation of this study that it tests the relationship of innovative proxies and countries' exports by cross-sectional data (allowing for a fixed time lag). Although the correlations between the two classes of high-technology capabilities of the OECD nations and their respective export performances are robust, this type of tests falls short of a proper account of the dynamics by which technical change determines international competitiveness (see Amendola and Perrucci 1992).[8] Bilateral comparison and analyses of the development of flows of trade within or between trade blocks show the extent to which the large industrialised nations have specialised in different product areas and how these priorities have developed over time. World-wide division of labour in the high-tech area is clearly visible (Legler et al., 1992).

The simple econometric tests cannot differentiate between the theories of comparative versus absolute advantages in external trade as both classical growth theory and evolutionary models provide similar expressions for empirical tests interpreting them in a different way. However, we observed that the comparative advantage concept seems to be verified for larger OECD countries on high-level markets in an ideal way, but is more doubtful for smaller developed countries like Finland or Australia trying to catch up in technology production. This points to increasing scale effects and would back the concepts of absolute advantages and virtuous circles more than the opposite assumption. Further clarifications are on the research agenda.

The analytical instruments introduced in this chapter seem to be well suited for broader application in economic analysis as well as in technology policy. Policies aiming at improvements in competitiveness ought to focus primarily upon the factors which influence international trade. *National* technology production seems to be a major component of market success up to now at least for the high-level technologies despite increasing intra-firm flows of knowledge.

However, one severe problem that remains is the lack of a concordance between the international trade (SITC III) and the international patent classification (IPC). In this chapter, the US Patent and Trademark Office concordance was used which cannot be transposed to the international classifications. The US patent system provides unique data sources in this respect and is more advanced than other countries or the OECD as a whole.

By applying the triad model and thus original statistical data from countries other than the United States, complicated classification transactions are needed which may create serious errors. In research efforts to follow, a patent-to-trade concordance on the international levels should be worked out. Case studies in information technology were recently performed (Grupp and Soete, 1993) which show convincingly that such an attempt can be successful, provided that economic and technological knowledge is present within the working group and includes detailed know-how on intellectual property rights.

Notes

1 Outstanding exceptions in this field are the evolutionary studies by Fagerberg (1988b) who analysed the contribution of technology gaps to domestic economic growth in a modified neo-Keynesian model, and Grossman and Helpman (1991) who integrated in their models the growth effects of technological leads due to different technological paths followed in each country. Both dealt explicitly with the impacts of technological and product specialisation on economic growth.

2 See, for instance, the literature cited in Dosi *et al.* (1990).

3 Krugman (1981) and Grossman and Helpman (1991) have developed formal models of what is called 'trade hysteresis' – that is the accumulation and path-dependence of changes in trade patterns.

4 The similarity in mathematical formulae does not allow for a differential econometric test in order to falsify some of the competitive theories or schools of economic thought.

5 Most interestingly, the German Bundesbank (see Deutsche Bundesbank, 1994), in a recent edition of its Monthly Report, published a macroeconomic test of several monetary indicators which are supposed to reflect the price competitiveness of countries in world trade. Contrary to common belief, neither the terms of trade nor the wage rate in industry proved successful in testing against German world market shares. Thus, the Bundesbank's economists concluded, if the concept of price competitiveness should be of any use at all, it is not determined by the relative price of labour costs. Almost in line with this argument is the study of Amendola and Perrucci (1992) who showed convincingly that technology and investment variables are more pertinent to explain trade dynamics than other factors. Their most general result concerns the *long-term* effect of technical change both in its disembodied form as captured by patents, and embodied into fixed investments upon export dynamics. Changes in wages and exchange rates, in Amendola's and Perrucci's econometric analysis, appear to affect only *short-run* changes in competitiveness which are reabsorbed in the longer term (Amendola and Perrucci, 1992, p. 463).

6 Israel, although not an OECD member state and not included in this chapter, would be the most striking case in this context, see Grupp *et al.* (1992).

7 Note that Belgium and Luxembourg which form an economic and monetary union are treated as a single country. Note also that all German data are for West Germany only. In the figures, the three-digit ISO country code is used. In alphabetical order: AUT (Austria), AUS (Australia), BELUX (Belgium and Luxembourg), CAN (Canada), CHE (Switzerland), DEU (Germany), DNK (Denmark), ESP (Spain), FIN (Finland), FRA (France), GBR (United Kingdom), ITA (Italy), JPN (Japan), NLD (Netherlands), NOR (Norway), SWE (Sweden) and USA.

8 One of the authors (G.M.) examines the dynamic relations in his doctoral thesis (Münt, 1996).

9 High-technology industries and international competition

PAOLO GUERRIERI AND CARLO MILANA

1 Introduction[1]

High-technology industries, which are usually defined as those with above-average expenditure on Research and Development (R&D) (OECD, 1985; Kodama, 1991), play a key role to long-term economic performance of all advanced economies. As a growing recent literature has shown, the importance of this group of sectors does not depend so much on the technological content of their products, but on the nature of their output which enters as technological inputs into a wide range of economic activities related to the production of goods and services. Over the past decade, competition among the major countries has strongly increased in high technology industries, and their competitive positions have been significantly altered. This chapter deals with these changes, by assessing the long-term trade performance of the most industrialised economies over the past two decades in high-technology industries. The chapter is in three sections. The next section deals with the main common features of high-technology industries and their special contributions to the performance of national economies; it also outlines the original taxonomy and data base for trade flows in high-technology products which are used in this chapter. Section 3 presents the empirical findings of our analysis with an evaluation of the changes in the competitive positions of the United States, Japan and the EU countries. In the final section, the chapter's main findings are summarised.

2 The main features of high-technology industries

High-technology product groups are related in that they embody, either directly, or indirectly through the intermediate goods used in their production, relatively intensive research and development inputs (Scherer, 1992). Other common features are equally important in shaping the competitive advantage of firms in the production and trade of high-technology goods (OECD, 1992a; Scherer, 1992): (i) the cumulative effect of innovative advantage, characterised by steep learning curves and significant dynamic economies of scale; (ii) the capability of generating positive external economies, in terms of hard-to-appropriate spillovers from one activity to another; (iii) strategic oligopolistic environments, in which small numbers of large interdependent companies compete through trade and transnational investment.

In industries with these characteristics, the relative advantage of one country vis-à-vis others stems not only from differences in national factor endowments, but as theory suggests and empirical evidence confirms, is also largely a function of differential technological knowledge and capability, which are created and reproduced through time[2].

During the last decade, high-technology industries have been the focus of special concern for the governments of all major countries (Ostry, 1990; Tyson, 1992). A variety of economic reasons are behind this concern: (i) high-technology industries account for large and growing shares of trade and investment in the industrialised area (see below); (ii) they are often the source of important technological innovations, the benefits of which are likely to spill over at intraindustry and interindustry levels (Griliches, 1990b; Mohnen, 1990; Scherer, 1982a); (iii) most of them are high-productivity industries and pay higher wages than do other manufacturing sectors (Katz and Summers, 1989).

Amongst these issues, that of the positive externalities generated by high-technology industries appears to be the most significant (Krugman, 1992; Chesnais, 1986). By definition externalities are benefits from market size effects, and pure informational spillovers which cannot be appropriated, priced and marketed. Furthermore, they are often locally and regionally concentrated, and benefits diffuse more rapidly from firm to firm and activity to activity within national borders than they do across borders (Krugman, 1991a; Porter, 1990). Many electronic industries are cases in point: their process and product innovations provide significative intraindustrial and interindustrial externalities which are strongly national and even local in character (Ernst and O'Connor, 1992; Mohnen, 1990).

This means that high-technology industries provide social benefits not reflected simply by the magnitude of their value added. The contribution of high-technology industries, because of this hidden contribution to national technological capability and economic welfare, clearly goes beyond their direct contribution to a country's production and trade balance. Thus the study of the evolution of these industries assumes particular importance. Even more so, since their role has been continuously growing over the past decade. In effect the shares of technology-intensive

Table 9.1. *Countries' shares in world trade in high-tech sectors*

	1970–2	1974–6	1977–9	1982–4	1985–7	1988–90	1990–2
OECD	95.8	93.7	91.0	86.9	85.6	82.7	79.9
United States	30.2	27.3	23.8	25.4	22.3	20.5	20.2
Canada	4.5	2.9	2.2	2.4	2.4	2.6	2.5
Japan	6.9	7.5	9.6	12.6	15.1	15.2	14.1
EU (12)	46.4	48.1	48.1	40.6	39.7	38.5	37.7
Germany	16.2	16.8	16.5	13.2	13.1	12.4	11.9
France	7.1	8.2	8.7	7.4	7.1	6.9	7.2
United Kingdom	10.3	9.4	9.9	8.5	7.6	7.6	7.5
Italy	4.5	4.2	4.1	3.7	3.7	3.4	3.1
Other EU countries	7.8	8.8	8.2	6.9	7.2	7.1	6.8
'Greece, Port., Spain'	0.5	0.7	0.7	0.9	1.0	1.2	1.2
EFTA	7.6	7.5	6.9	5.5	5.9	3.7	5.2
Non-OECD	3.7	5.9	8.0	11.8	13.5	15.8	18.1
Asian NICs	1.0	2.3	3.3	5.8	7.6	9.7	10.9
American NICs	0.6	0.7	0.8	1.3	1.6	1.2	1.0
Eastern Europe	0.6	0.6	0.6	0.4	0.3	0.3	0.3

Notes:
Percentage shares in values; average values in each sub-period.
Asian NICs : Hong Kong, Singapore, South Korea, Taiwan.
American NICs : Argentina, Brazil, Mexico.
Source: SIE World Trade data base.

production and trade in total manufactured output and export have increased steadily during the past two decades. In fact, the growth in world high-technology exports, particularly in electronic goods, has been consistently higher than the average growth in world manufactured exports during the 1980s (Guerrieri and Milana, 1991).

It should also be noted that such trade is highly concentrated in the most industrialised countries. In 1990, almost 75 per cent of the world's high-technology exports were located in the United States, Japan and the EU countries (table 9.1). Over the past two decades competition among these countries and areas has strongly increased, and significant changes in their relative competitive positions have taken place. The purpose of the next section is to assess these changes, by analysing the trade performances in high-technology products of all major countries. For this end we use an original taxonomy and trade data base.

As to the taxonomy, there are different and alternative methodologies to define and quantify 'high-technology trade'. They can be grouped into two broad sets of measures. The first measure uses an indicator of technological inputs, such as R&D to sales ratio or engineers and scientific personnel to regular employees ratio as a proxy for embodied technology, and defines as 'high-technology' industries all those sectors characterised by a ratio higher than a given threshold value. All traded products included in these industries are then classified as high technology products (International Trade Administration, 1985; OECD, 1985).

The second approach uses more detailed product data and relies upon the evaluation of industry experts in order to determine the technological content of the various products. At this level of detail, analysts' judgements are used to determine whether a product is high technology (Abbott, 1991).

There is no doubt that the two methodologies for the classification of high-technology products are different in their approaches, and that the usefulness of either measures depends upon the particular application (Abbott, 1991). Both methods, however, suffer from several major flaws. The first method takes objective criteria to evaluate industry embodied technology (like R&D to sales), yet the high level of aggregation of industries leads to the assumption that all of the products comprising a so-defined high-technology industry are high-tech. This is clearly not true given the different technological content of the products included in a given industry. The second method provides a more accurate list of individual high-technology products, yet it is entirely subjective, with no quantifiable criteria by which the industry analysts make out this list.

In this study an intermediate classification system of the two alternative methods mentioned above is used (Guerrieri and Milana, 1991). According to the first approach, the R&D to sales ratio as a proxy for measuring different industry technological intensities is used,[3] and 'high-tech' industries have been defined as those industrial sectors characterised by a R&D to sales ratio higher than a given threshold value (4 per cent), thus following the OECD (1985). In order to overcome the drawbacks stemming from the high level of aggregation of the first approach,[4] a given number of traded products defined at the high level of disaggregation of five-digit SITC is then associated to each high-technology industry (Standard International Trade Classification). To this end, a correspondence between the International Standard Industrial Classification (ISIC) and the trade classification (SITC) has been established.

According to the second method of classifying high-technology products, a group of experts knowledgeable in each industrial sector, drawn from industrial associations, academic institutions, research department, as well as from firms, were asked to examine the individual products associated with high-technology industries in order to determine their real technological content and establish which products are to be considered high-tech. This procedure has allowed to set up a list of high-technology products.[5] Finally, data on imports and exports of high-technology products were aggregated by using an original trade data base (SIE World Trade Data Base) derived from the OECD as well as United Nations trade statistics (Guerrieri and Milana, 1990; Guerrieri, 1992, 1993).[6] Its main advantage lies in the fact that it allows for analysis of extremely disaggregated time series, thanks to its system of correspondence between the SITC Revised, the SITC Revision 2 and the SITC Revision 3.

3 Changes in the world trade of high-technology products

This section analyses the trade performance of the most advanced countries in high-technology products from 1970 to 1992 with the aim of identifying long-term changes in their competitive positions occurring in this period. For this end a variety of indicators are used because no single indicator can provide an adequate view of a country's international trade performance.

In the first two decades following the Second World War, US firms enjoyed almost unchallenged supremacy across a broad set of high-technology goods (Nelson, 1990c). An overview of our evidence reveals that over the period here considered there was a move from the technological hegemony of the United States to an oligopolistic structure in which several countries have the capability to affect the future direction of technological change.

The significant decline in the US relative competitive position in high-technology trade is reflected by its decreasing market share (table 9.1) and standardised trade balance, which is an indicator of relative competitive positions,[7] over the past two decades (figure 9.1). Certainly, the sharp deterioration of both, and in particular of the US trade balance during the 1980s, was to a certain extent the result of the unsustainably high value of the US dollar relative to other major currencies in the first half of the 1980s. Yet, the dollar's devaluation in the second half of the 1980s has significantly improved only the trade surplus with respect to the EU countries, as the huge trade deficit towards Japan remained almost stable; and the negative trade balance with respect to the Asian NICs (Hong Kong, Singapore, South Korea and Taiwan) only marginally decreased (figure 9.2). In contrast, the dollar's drop has produced big gains in the trade of many low-technology manufactures (Guerrieri, 1992; Krugman, 1992).

Further evidence of the US deteriorating competitive position derives from an application of the 'Constant Market Shares Analysis' (CMSA) to the observed changes of US export share in high-technology products, as it makes it possible to distinguish the effects of factors which determine competitiveness from those connected to structural changes (Magee, 1975). The 'structural effect' refers to the geographic and commodity structure of

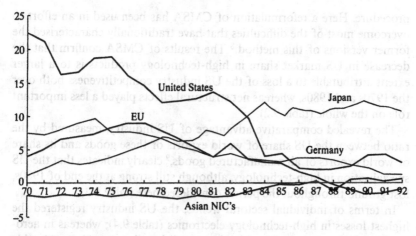

Figure 9.1 Trade Balance in High-Tech Products (as percentage of total trade in high-tech products
Note: * As percentage of total trade in high-tech products

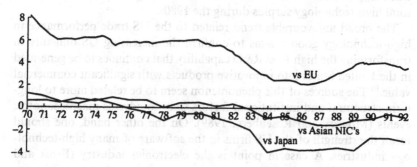

Figure 9.2 Bilateral Trade Balance in High-Tech Products of the US (as percentage of total world trade in high-tech products)
Note: * As percentage of total trade in high-tech products

a country's exports relative to the structure and the dynamics of world demand. They are positive (negative) if a country concentrates its exports on markets and/or commodities that grow faster (slower) than the world average (world demand). The competitiveness effect reflects the actual changes of a country's market share, assuming that its trade structure is constant; it represents that part of a country's trade performance deriving from its competitive factors, both 'price' and 'non-price' ones. Although the structural effect and the competitive effect may be interdependent from the causality point of view, they can be separated by means of an accounting procedure. Here a reformulation of CMSA has been used in an effort to overcome most of the difficulties that have traditionally characterised the former versions of this method.[8] The results of CMSA confirm that the decrease in US market share in high-technology products is to a larger extent attributable to a loss of the US industry competitiveness both over the 1970s and 1980s, whereas net structural effects played a less important role on the whole (table 9.2).

The revealed comparative advantage of US industry, measured by the ratio between the US share of world exports of these goods and its share of world exports of all manufactured goods,[9] clearly indicates that the US specialisation in high-technology, although still strong at the end of 1980s, lost ground throughout the period considered here (table 9.3).

In terms of individual sectoral trends, the US industry registered the highest losses in high-technology electronics (table 9.4); whereas in aerospace and related products, which have traditionally received considerable public sector support, the United States showed by the late 1980s a large market share – more than 40 per cent of world trade – and a substantial trade surplus (table 9.7), which amounted on average to almost half the US total high-technology surplus during the 1980s.

The broad unfavourable trend related to the US trade performance in high-technology goods seems to confirm the increasing US difficulty in transforming the high-level R&D capability that continues to be generated in the United States into innovative products with significant commercial value.[10] The sources of this phenomenon seem to be related more to long-term structural and institutional factors than to adverse macroeconomic events (Mowery and Rosenberg, 1989). On the other hand, one should recall the strength of the US firms in the software of many high-technology industries. A case in point is the electronics industry (Ernst and O'Connor, 1992).

Over the past two decades Japan is the country that has gained most in high-technology trade, as shown by the truly remarkable improvement of all trade performance indicators analysed here (market share and standardised trade balance) (table 9.1 and Figure 9.1). Figure 9.3 shows that during the 1980s, Japan had highly positive trade balances with respect to the two major areas (US and EC) and also to the East Asian NICs (Hong Kong, Singapore, South Korea, Taiwan). No other major country registered a similar positive trend.

The substantial improvement in the Japanese competitive position has been mainly concentrated in three major high-technology groups: electronics, mechanicals and, to a lesser extent, scientific instruments (tables 9.4–to 9.8). Particularly in the electronics industry (hardware), Japanese firms have gradually assumed a strong competitive position in world trade,

Table 9.2. Results of constant market shares analysis of the exports in high tech products, 1973–1990 (percentage values)

	Changes in market shares	Competitiveness effect	Structural effect			
	(c)=(d)+(e)	(d)	Total (e)=(f)+(g)+(h)	Market effect (f)	Commodity effect (g)	Specific effect (h)
Canada	−0.84	−1.02	0.19	0.78	2.03	−2.62
United States	−4.41	−2.89	−1.53	−2.00	0.01	0.46
Japan	5.33	1.58	3.75	2.95	0.61	0.18
Asian NICs	7.50	2.33	5.17	2.00	3.65	−0.48
Germany	−5.42	−3.55	−1.88	−2.10	−1.02	1.24
France	−0.32	−0.24	−0.08	−0.86	0.65	0.12
United Kingdom	−2.34	−1.84	−0.50	−0.56	0.47	−0.42
Italy	−0.78	−0.38	−0.40	−0.60	−0.04	0.25

Table 9.3. *Revealed comparative advantage of selected countries in high tech products*

	United States		Japan		Asian NICs		EU (12)	
	1970–3	1990–2	1970–3	1990–2	1970–3	1990–2	1970–3	1990–2
Total high-tech	219	178	80	123	54	112	99	89
Chemicals and drugs	111	116	86	44	45	66	123	126
Mechanicals	156	129	93	142	21	65	108	97
Electronics	212	146	110	189	132	194	95	66
Aerospace	440	384	6	7	16	13	63	98
Scientific instruments	207	198	86	100	15	41	103	105

	Germany		France		United Kingdom		Italy	
	1970–3	1990–2	1970–3	1990–2	1970–3	1990–2	1970–3	1990–2
Total high-tech	111	88	97	107	132	131	79	62
Chemicals and drugs	159	123	99	151	103	116	114	76
Mechanicals	140	124	105	120	141	113	79	61
Electronics	99	51	97	71	113	116	88	49
Aerospace	20	78	87	180	175	186	41	58
Scientific instruments	138	126	93	94	135	177	53	55

Notes:
*Ratio of a given country's share in world export in high tech products to its share in world trade in manufactures.

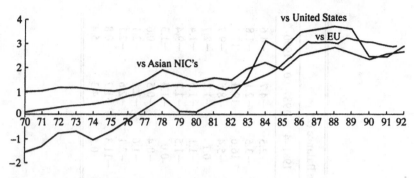

Figure 9.3 Bilateral Trade Balance in High-Tech Products of Japan

achieving the most significant gains over the 1980s, in spite of the revaluation of the yen in terms of the dollar during the second half of the past decade (table 9.4).

The results of CMSA (table 9.2) show that both positive competitiveness and especially structural effects contributed to the remarkable gains in export shares achieved by the Japanese industry. Japan, after having suffered unfavourable structural changes in the 1970s particularly in terms of a negative 'product composition' effect, has shifted its export flows towards more dynamic markets and products in the 1980s. On the other hand, the main explanations of the rise of Japan in high-technology production stress either factors at firm-level, in particular matters of internal organisation and management, or factors at industry level associated with the profound transformations in the specialisation pattern of Japanese industry over the past 15 years (Freeman, 1987; Kodama, 1990; Guerrieri, 1992; Okimoto 1989). Within this pattern, high-technology has played a key role, as confirmed by the huge rise in Japan's revealed comparative advantage in total high-technology products and in two high-technology industries in particular (electronics and mechanicals, see table 9.3). In this regard, one should add – as many pointed out – that these favourable trends have been at least partly generated by government sector-specific policies and by the advantages of a quasi-closed domestic market, created by 'peculiar barriers' to market access (Johnson, Tyson and Zysman, eds., 1990; Kreinin, 1993; Lawrence, 1991), thus enhancing trade conflicts with the United States and Europe.

In addition to Japan, it is important to note the remarkable performances of East Asian NICs (Hong Kong, Singapore, South Korea, Taiwan) in high-technology trade over the entire period (1970–90), although almost entirely concentrated in electronics. As shown in table 9.1, their market shares, while negligible in the early 1970s, have increased to

Table 9.4. *Shares and trade balance in high-tech electronics*

	Market Shares (1)					Trade Balance (2)				
	1970–3*	1976–9	1981–4	1987–90	1990–2	1970–3	1976–9	1981–4	1987–90	1990–2
United States	28.8	23.2	25.1	17.8	17.1	16.7	8.1	2.5	−4.7	−4.6
Canada	2.9	2.0	2.4	2.0	2.1	−3.7	−2.9	−3.5	−2.1	−1.8
Japan	9.6	14.6	20.0	23.0	21.3	4.1	10.6	16.6	19.4	17.1
EU (12)	45.2	41.9	32.6	30.0	28.0	9.9	−2.4	−5.4	−8.5	−9.7
Germany	14.8	12.9	9.0	7.7	7.0	3.2	1.5	0.1	−1.2	−2.7
France	7.3	7.0	5.4	4.7	4.3	−1.4	−0.7	−1.1	−1.5	−1.4
United Kingdom	8.6	8.1	7.1	6.9	6.6	−0.8	−0.4	−1.5	−1.7	−9.4
Italy	5.1	3.9	3.1	3.0	2.6	0.4	−0.8	0.0	−1.4	−1.3
Other EU Countries	8.8	9.3	7.3	7.1	6.8	−1.3	−0.2	−0.4	−0.5	−0.6
Greece, Port.,Spain	0.6	0.6	0.7	0.7	0.8	−2.2	−1.9	−1.6	−2.1	−2.1
EFTA	6.7	5.6	3.9	3.6	3.4	−1.4	−1.9	−2.1	−2.6	−2.1
Non-OECD	5.2	11.3	15.6	22.2	26.8	−13.1	−13.4	−11.3	−1.7	−0.8
Asian NICs	3.1	7.8	10.3	16.1	18.8	−0.1	0.1	0.5	4.3	5.4

Notes:
(1) Ratio of a country's or area's exports to total world exports in high-tech electronics.
(2) As percentage of total world trade in high-tech electronics.
* average values in each sub-period.
Source: SIE World Trade data base.

Table 9.5. *Shares and trade balance in world trade in high-tech chemicals*

	Market share (1)					Trade balance (2)				
	1970–3*	1976–9	1981–4	1987–90	1990–2	1970–3	1976–9	1981–4	1987–90	1990–2
United States	15.0	13.8	16.9	13.0	12.8	8.2	6.4	8.4	5.5	5.6
Canada	1.0	1.7	2.1	2.3	2.1	–3.6	–1.4	–0.5	–0.3	–0.4
Japan	8.0	5.8	5.7	5.4	5.3	3.9	1.9	0.1	0.3	0.5
EU (12)	58.2	59.3	53.5	56.0	52.7	22.2	12.4	9.1	6.9	5.4
Germany	23.9	20.4	17.1	18.4	17.1	14.6	9.4	6.5	7.5	6.6
France	7.2	10.2	10.6	10.3	10.1	–1.8	–0.5	0.2	0.5	0.9
United Kingdom	8.0	8.8	7.2	6.9	6.5	2.6	2.8	1.0	–0.3	–0.1
Italy	6.3	5.6	4.7	4.5	4.0	0.7	0.1	–1.0	–2.6	–3.0
Other EU Countries	12.2	13.5	12.7	13.9	13.0	1.6	3.2	4.1	4.2	3.3
Greece, Port., Spain	0.5	0.8	1.3	1.9	2.0	–4.1	–2.7	–1.7	–2.3	–2.5
EFTA	11.7	10.3	9.4	9.1	8.7	2.2	1.7	1.5	1.3	1.6
Non-OECD	5.4	8.0	10.6	13.1	16.3	–25.8	–22.1	–19.8	–15.5	–12.1
Asian NICs	1.1	1.5	2.6	4.4	6.4	–22.9	–2.8	–2.4	–2.8	–1.5

Notes:
(1) Ratio of a country's or area's exports to total world exports in high-tech chemicals and drugs.
(2) As percentage of total world trade in high-tech chemicals and drugs.
* average values in each sub-period.
Source: SIE World Trade dasta base.

Table 9.6. *Shares and trade balance in high-tech mechanicals*

	Market Shares (1)					Trade Balance (2)				
	1970–3*	1976–9	1981–4	1987–90	1990–2	1970–3	1976–9	1981–4	1987–90	1990–2
United States	20.9	19.1	20.0	15.7	15.3	6.1	6.8	4.4	-4.1	-1.2
Canada	5.9	3.1	2.2	3.1	3.0	-4.2	-4.3	-4.4	-3.4	-3.1
Japan	8.2	11.7	14.8	16.8	16.0	6.3	9.9	12.6	14.3	13.2
EU (12)	51.8	51.6	43.9	42.3	42.2	25.7	19.5	14.9	7.5	7.6
Germany	21.1	21.1	17.2	18.2	17.6	14.2	13.9	10.3	9.2	7.9
France	7.7	9.6	8.0	7.8	8.0	1.8	3.9	3.1	2.5	2.7
United Kingdom	11.0	9.6	8.1	6.3	6.5	6.6	5.3	3.1	0.1	0.7
Italy	4.4	4.2	4.4	3.7	3.5	0.2	0.6	1.9	-0.2	-0.2
Other EU Countries	6.8	6.1	4.9	4.7	4.7	-4.1	-2.9	-1.8	-2.2	-1.7
Greece, Port.,Spain	0.8	1.0	1.3	1.6	1.8	-2.3	-1.3	-0.9	-1.8	-1.9
EFTA	9.3	8.3	7.4	8.4	8.1	-0.2	0.7	1.4	1.7	1.8
Non–OECD	3.6	5.9	11.0	12.9	14.6	-27.6	-34.9	-30.5	-16.2	-15.5
Asian NICs	0.5	1.4	3.9	5.8	6.4	-2.9	-3.1	-2.5	-2.9	-3.3

Notes:

(1) Ratio of a country's or area's exports to total world exports in high-tech mechanicals.

(2) As percentage of total world trade in high-tech mechanicals.

* average values in each sub-period.

Source: SIE World Trade data base.

Table 9.7. *Shares and trade balance in world trade in high-tech aerospace*

	Market Shares (1)					Trade Balance (2)				
	1970–3*	1976–9	1981–4	1987–90	1990–2	1970–3	1976–9	1981–4	1987–90	1990–2
United States	58.7	50.4	42.0	43.3	40.1	51.1	44.6	29.7	28.9	27.9
Canada	7.6	3.3	3.2	3.8	3.4	1.8	−0.5	−1.6	−0.8	0.4
Japan	0.6	0.3	0.5	0.8	0.8	−5.5	−3.0	−4.1	−4.0	−3.3
EU (12)	29.6	37.8	41.6	40.1	42.7	2.1	1.8	4.9	3.2	3.3
Germany	3.0	8.0	12.4	11.1	11.2	−2.9	−1.7	−1.5	−1.1	−1.9
France	6.5	9.0	8.4	10.0	12.0	−0.3	3.0	3.2	3.3	4.3
United Kingdom	13.2	13.9	13.2	10.4	10.8	4.8	3.0	4.7	3.1	3.2
Italy	2.2	3.0	3.8	3.6	3.4	−0.9	0.8	1.3	0.5	0.1
Other EU Countries	4.7	3.6	3.2	3.8	4.1	−5.1	−2.3	−1.3	−1.4	−0.8
Greece, Port., Spain	0.1	0.3	0.6	1.2	1.2	−2.9	−1.0	−0.8	−1.1	−1.5
EFTA	1.0	0.9	0.8	1.5	1.6	−3.3	−2.6	−2.9	−2.4	−1.9
Non-OECD	2.2	5.0	5.4	4.5	3.6	−18.3	−24.3	−16.4	−13.7	−14.3
Asian NICs	0.4	1.3	1.4	1.4	1.2	−1.7	−2.7	−2.9	−3.89	−4.3

Notes:
(1) Ratio of a country's or area's exports to total world exports in high-tech aerospace products.
(2) As percentage of total world trade in high-tech aerospace products.
* average values in each sub-period.
Source: SIE World Trade data base.

Table 9.8 *Shares and trade balance in high-tech instruments*

	Market Shares (1)					Trade Balance (2)				
	1970-3*	1976-9	1981-4	1987-90	1990-2	1970-3	1976-9	1981-4	1987-90	1990-2
United States	28.0	25.6	29.7	22.3	22.8	20.8	18.2	19.8	9.8	10.9
Canada	3.3	2.0	2.5	1.9	1.7	-4.0	-2.8	-2.4	-2.5	-2.7
Japan	7.7	7.8	9.0	11.9	11.9	2.5	3.5	4.0	6.1	6.2
EU (12)	48.5	49.3	43.1	46.3	44.5	14.5	6.2	4.4	3.1	1.5
Germany	20.7	19.6	16.2	19.0	17.8	11.6	9.7	7.6	9.3	7.5
France	6.9	8.4	6.6	6.4	6.3	-2.1	-0.2	0.0	-0.8	-0.9
United Kingdom	10.4	10.2	10.7	11.1	10.2	3.0	2.1	1.3	1.8	1.6
Italy	3.0	2.9	2.8	3.1	3.2	-2.7	-1.7	-0.7	-3.1	-2.8
Other EU Countries	7.3	7.8	6.2	6.2	6.1	-3.4	-1.2	-0.5	-1.1	-0.9
Greece, Port.,Spain	0.3	0.4	0.5	0.6	0.9	-3.1	-2.5	-2.2	-3.0	-3.0
EFTA	9.4	10.5	9.8	9.6	9.3	1.0	1.3	1.9	1.4	1.6
Non-OECD	2.7	4.4	5.4	6.9	8.8	-22.5	-27.2	-26.1	-20.4	-19.2
Asian NICs	0.4	1.3	2.0	3.7	4.0	-1.8	-2.5	-3.4	-4.7	-5.6

Notes:
(1) Ratio of a country's or area's exports to total world exports in high-tech professional instruments.
(2) As percentage of total world trade in high-tech professional instruments.
*average values in each sub-period.
Source: SIE World Trade data base

account for approximately 9.7 per cent of the world market. The CMSA reveals that the increase in the aggregate export share was largely made of a positive structural effect which has turned out to be by far more important than the 'competitive' effect during the whole period under consideration (table 9.2). The ability to shift the export composition towards highly dynamic markets (and, especially, particular commodities within these markets has played the major role in improving the competitive position of Asian NICs, although improvement in competitiveness has not been negligible. Trade balances in high-technology industries in Asian NICs, unlike those in the most advanced areas, have registered persistent deficits over most of the period examined here. Trends in the relative competitiveness indicator (standardised trade balance), however, show a significant reversal of this pattern beginning in the early eighties (figure 9.1).[11] Over the second half of the 1980s, Asian NICs have registered trade surpluses with respect to both the US and the EU countries. This, however, has only partially compensated for their growing deficit in relation to Japan (figure 9.3).

Change in the export patterns of Asian NICs has also been significant. The specialisation index clearly reveals an improved comparative advantage for Asian NICs in the high-technology group as a whole although entirely concentrated in electronics (table 9.3). This trend stems from the diversification process of manufacturing output since the second half of the 1970s (Bradford and Branson, 1987; Chow and Kellman, 1993). The development strategies of the Asian NICs were initially based on competitiveness 'clusters' including labour-intensive consumer goods exports and diversification of industrial structure has gradually been carried out through a strengthening of capital-intensive productions, and particularly of technology-intensive industries. Within the latter group, the significant achievements of the Asian NICs in many electronic sectors is emblematic (Ernst and O'Connor, 1992) (see table 9.4). To this end either state interventions or incentives and subsidies policies were used on a large scale and in different forms (Amsden, 1989; Wade, 1990).

Finally, the EU countries suffered significant reductions in their world market shares and increasing trade deficits during the 1980s (table 9.1, figure 9.1). These negative trends affected all EU major countries, including Germany, which confirms its weakness in this area (Guerrieri, 1993; Legler, 1990). By the late 1980s, out of the three major regions, only the EU suffered high trade deficits (figure 9.4). In terms of trade structure, the EU specialisation in high-technology trade, as shown by the measure of 'revealed comparative advantage' (the ratio of their export shares in that sector to all manufactured exports) and that of Germany in particular, decreased significantly (table 9.3), displaying values even below 100. Again the evidence suggests that high-technology goods have played and continue

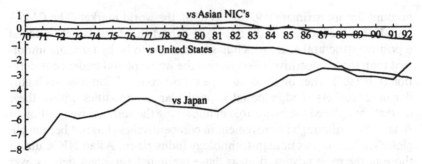

Figure 9.4 Bilateral Trade Balance in High-Tech Products of the EU

to play a relatively minor role in the trade structure of the EU countries (with the exception of the UK) compared to that of the US and Japan.

Individual trends of the main high-technology product categories would seem to suggest that the competitive position of EU industry is not as unfavourable as noted above. Sectoral evidence shows a consolidation of European competitiveness in chemicals and pharmaceuticals, a relative holding pattern in scientific and professional instruments, and, more recently, a rise in aerospace (tables 9.5, 9.7 and 9.8). In contrast to these positive trends, however, the EU countries lag behind in the electronics industry (table 9.4). During the 1980s, the EC countries suffered significant reductions in their world shares and increasing trade deficits, primarily with the US and Japan and, to a lesser extent, with the Asian NICs. Given the prominent role played by electronics technology and products in the current restructuring of industries of all major countries, concerns over the relative weakness of EU producers are even more worrisome (Freeman *et al.*, eds., 1991; Patel and Pavitt, 1991b). The sources of this disappointing performance of the EU countries are many: excessive market fragmentation (Commission of the EC, 1988; inadequate and inefficient R&D (Sharp and Shearman, 1987), weak national industrial policies with respect to problems associated with current global competition (Pavitt and Sharp, 1993) and failure of internal organisation and management at the firm-level (Amable and Boyer, 1992). In this regard, some progress has recently been made with the completion of a unified European market. But it represents only a first step in the right direction, which is not able per se to offset the long-term trend of gradual erosion of EU competitiveness in high-technology (Freeman *et al.*, eds., 1991).

4 Conclusions

This chapter has considered world trade competition in high-technology products. The special contribution to a nation's long term economic performance of high-technology industries is related to the fact that they generate spillovers and externalities effects. The remainder of the chapter has been concerned with the statistical evidence surrounding the trends in high-technology trade over the past two decades. By using an original data base and taxonomy for trade flows it has been shown that in the period from 1970 to 1992, there were significant changes in the competitive positions of the United States, the EU countries and Japan in high-technology products. First there was a move from the technological hegemony of the United States to an oligopolistic structure in which several countries have the capability of affecting the future direction of technological change. Within this new competitive environment Japan undoubtedly achieved the highest gains, whereas the United States' competitiveness has experienced a process of relative deterioration. The EU industry, in particular, has suffered sharp losses on domestic and international markets, particularly in electronics where the Asian NICs emerged as new strong competitors. This general evolution, however, has been sharply differentiated with respect to the various product groups included in the high-technology category.

Finally, the empirical evidence of the chapter seems to suggest that these changes in competitiveness in high-technology trade constitute long-term competitive trends. They should be attributed to structural factors besides macroeconomic and exchange rate fluctuations. Given the special concern national governments place on competition in high-technology industries, trade friction and new forms of public intervention in attempts to correct these imbalances are bound to proliferate in the industrialised area in the near future.

Notes

1 This research has been supported by the Italian Consiglio Nazionale delle Ricerche within the project 'Technological change and industrial growth'.
2 In this regard the literature is extremely vast. Emblematic contributions are Dosi, Pavitt and Soete (1990) and Grossman and Helpman (1991).
3 Although the drawbacks of the ratio of R&D expenditures to sales as an indicator of technology intensity are well known, yet for a comparative analysis comprising a large number of heterogeneous countries like ours, no other objective measure as reliable as R&D international statistics is so far available.
4 Care was taken to ensure that the mix of experts would not be biased towards any specific sector or technological field. One was sometimes able to confirm independently the information provided by the experts and the firms.
5 Classification of high-technology products for this chapter comprises the following list of product groups (SITC 5 digit level):

  1 Chemicals and drugs: Synthetic organic dyestuffs; radio-active and associated materials; polymerisation and copolymerisation products (polyethylene, polypropylene, polystyrene, polyvinyl); Antibiotics

2 Mechanicals: Turbines; electric motors; electric power machinery and apparatus; internal combustion piston engines
3 Electronics: Automatic data processing machines and Units; telecommunications equipment; semi-conductor devices; electronic microcircuits
4 Scientific and professional instruments: Electronic measuring instruments; medical instruments; optical instruments and appliances; photographic apparatus and equipment
5 Aerospace: Aircraft and associated equipment; Spacecraft.

6 The SIE-World Trade data base provides detailed information on the exports and imports of 83 countries with respect to 400 product groups. The data base includes trade statistics with respect to the 24 OECD countries, the newly industrialising countries (NICs), the other developing countries and the former CMEA countries, and makes it possible to examine and analyse the entire world trade matrix. The source for the basic trade statistics of the SIE-World Trade is the official publications of the OECD and the United Nations provided on magnetic tapes. The SIE data base is organised in different product group classifications at various levels of disaggregation (400 product groups, 98 sectors, 25 categories, five branches) according to the three Standard International Trade Classifications (SITC), Revised, Revision 2, Revision 3, defined by the Statistical Office of the UN as to the periods 1961–75, 1978–87, 1988 onwards.

7 The standardised trade balance or the indicator of relative competitive position (IRCP) highlights the international distribution over time of trade surpluses and deficits among countries in each group of products. Trade surpluses and deficits are standardised by total world trade in the same group of products (CEPII, 1983, CEPII, 1989). The evolution of trade balance distribution reveals competitiveness patterns of various countries in a certain group of products. For each country (j) and product group (i) the indicator is given by:

$$\text{STB or IRCP} = \frac{X_i - M_i}{WT_i}$$

X_i=total exports of country (j) in the product group (i); M_i=total imports of country (j) in the product group (i); WT_i=total world trade in the product group (i).

8 The version of the CMSA applied in this chapter decomposes a change in a country's export share into the following four effects: (a) competitiveness effect: it measures the change of a country's export share due only to competitiveness factors assuming that its trade structure (market and commodity) is constant; (b) market effect: it represents the influence of the geographic composition of trade flows upon the aggregate export share of a country. It is positive (negative) if a country concentrates its exports on markets that grow faster (more slowly) than the world average; (c) commodity effect: it represents the influence of the product composition of trade flows upon the aggregate export share of a country. It is positive (negative) if a country concentrates its exports on products for which demand is growing faster (more slowly) than the world average; (d) specific market-commodity effect: it represents the influence on the aggre-

gate export share of a country stemming from specific composition product markets more (or less) favourable.

The sum of (b), (c) and (d) effects represents the overall 'structural effect', which measures those changes in aggregate export share of a country due only to changes in commodity-market structure in world trade. For further details on the methodologies of CMSA here used see Milana (1988) and Guerrieri and Milana (1990).

9   This is the ratio of the share of individual countries in world exports (imports) of a given product group to the share of the same country in total manufacturers world exports (imports).

10 See Scherer (1992).

11 They have continued to improve in recent years, achieving significant positive trade balances in the late 1980s in the cases of Singapore and Taiwan.

10 User–producer interaction, learning and comparative advantage

JAN FAGERBERG

1 Introduction

Michael Porter's book *The Competitive Advantage of Nations* (1990) has led to increasing attention to the favourable impact that the domestic market, through 'advanced domestic users', may have on the international competitiveness of a country. The idea that the domestic market may affect competitiveness positively, is by no means a new one: it dates back at least to List (1841). However, neo-classical trade theorists have normally regarded it as 'theoretically unsound' and as a cover for protectionism. This chapter presents a critical appraisal of the theoretical and empirical evidence on this hypothesis. Based on an empirical method initially developed by Anderson *et al.* (1981), the hypothesis – of a positive impact of the domestic market on the competitiveness of a country – is tested on data for 16 OECD countries between 1965 and 1987.

2 Why should 'advanced domestic users' matter?

Traditionally, most attempts to explain the specialisation patterns of countries in international trade have focussed on supply conditions. According to standard neo-classical theory of international trade, countries ought to specialise in areas of production that make intensive use of factors of production with which the country is relatively well equipped. In spite of the dominant role played by traditional neo-classical theory in this area, there has always been a strand of thought that has emphasised learning as a potential source of comparative advantage. This tradition points to the potential effects of relations between firms or sectors, within the domestic economy, on innovation and learning, and the impact of this on the international competitiveness of the country and its specialisation pattern in international trade.[1]

The first systematic attempt to discuss the implications of these ideas for trade theory was made by Linder (1961). His argument runs as follows. First, a need that cannot be sufficiently satisfied by existing products arises on the demand side. Since entrepreneurs for various reasons (culture, language, proximity) tend to be better informed about developments in the home market than in markets elsewhere, they will usually be the first to react to the demand for new or improved products arising in the domestic market. The outcome of this activity, i.e., the innovation, then enters a period of testing and revision in which the home market is assumed to play a critical role. If the new product is a success at home, it will probably be introduced on the export market too. Thus, in the case of developed countries, he suggested that it is demand-induced innovation *within* each *country*, not supply factors, that determines comparative advantage.

Recently, Porter (1990) has presented an evolutionary scheme of economic development based on similar ideas. Echoing Linder, he argues that traditional supply factors, although important in the earlier stages of development, are not among the prime determinants of 'competitive advantage' in more advanced countries, where growth is assumed to be innovation driven. The most competitive industries in an advanced country, he argues, tend to be highly integrated ('clustered'), both vertically and horizontally, with favourable consequences for learning, innovation and 'competitive advantage'. In Porter's scheme, this typically starts with integration between customers in traditional industries and suppliers of machinery and other types of advanced equipment, then widens through spillovers and feedbacks to and from related and supporting industries (Porter, 1990, pp. 554–5).

In this chapter, we will focus on the first of these two mechanisms, emphasised by both Linder and Porter, i.e., that a high degree of integration between customers and suppliers (or users and producers) may affect international competitiveness/comparative advantage positively. This hypothesis is intuitively appealing, and there is a large amount of descriptive evidence that can be used in its defence (see Porter, e.g., 1990). However, in spite of the growing popularity of this approach, many still probably feel that it is a phenomenon in search of a theory. We will briefly sketch a possible framework for the analysis.

Let us assume that the development of new technology in many cases requires close communication and interaction between users and producers of technology (Lundvall, 1985, 1988). To achieve this end, a channel – and a common code – of communication must exist. The establishment of channels and codes of communication involves fixed costs, and this implies that in a stable user–producer relationship, the cost per transaction is decreasing. This is clearly an argument for keeping relationships stable. Furthermore, lower transaction costs are likely to lead to a higher volume of transactions. Hence, a higher rate of innovation should be expected in a market characterised by enduring user–producer relationships, compared with a more 'atomistic' market structure. To the extent that the parties of a stable user–producer relationship can prevent the (immediate) diffusion to others of the innovations they make, as seems likely, they may (for some time at least) keep the benefits for themselves.[2] Indeed, the fact that the relationship is of an enduring character, and is recognised as such by both parties, may significantly increase the probability of appropriating the benefits. Thus, a stable user–producer relationship may be interpreted as an institution that reduces the costs – and increases the pace – of innovation and learning, while at the same time making it easier to appropriate the economic benefits.[3] As a result, the competitive positions of the participating firms are likely to improve. To some extent this holds for both users and producers, but in this chapter we shall focus mainly on the latter.

However, the importance of stable relationships may vary across industries. It should be expected to be of special importance in industries characterised by complex and user-specific technology. In these cases, the need for close communications and interaction between users and producers of technology is likely to be large, and the costs of establishing new relationships of this kind high. In other cases, products and technologies may be highly standardised: both transaction costs and the need for enduring user–producer relationships will be low.

To the extent that this type of interaction takes place mainly *within country borders*, this should be expected to affect patterns of export specialisation (or competitive advantage) of countries as well. Since, as pointed out by both Linder and Porter, the costs associated with communication and interaction increase with distance and differences in culture, language, institutional settings, etc., this may be a reasonable assumption to make. Porter even holds that the importance of the domestic market for competitive advantage is growing.

While globalization of competition might appear to make the nation less important, instead it seems to make it more so. With fewer impediments to trade to shelter uncompetitive domestic firms and industries, the home nation takes on growing significance because it is the source of the skills and technology that underpin competitive advantage. (Porter, 1990, p. 19)

However, it may also be argued that the increasing role of multinationals in world production has reduced the costs of communication and interaction significantly, and that the Linder–Porter hypothesis therefore was more relevant in the past than it is presently. The empirical evidence presented in this chapter may shed some light on this controversy.

Another issue raised by Porter is to what extent a competitive market structure is a necessary condition for a positive impact of user–producer interaction on competitiveness. He argues that 'favourable demand conditions . . . will not lead to competitive advantage unless the state of rivalry is sufficient to cause firms to respond to them' (1990, p. 72). Porter seems to be most concerned with competition among producers (suppliers) of technology.[4] However, a similar argument holds for the user side: users that are under continuous pressure to improve their performance, are more likely than others to demand improvements from their suppliers. Here, Porter especially emphasises the importance of international competition:

One competitive industry helps to create another in a mutually reinforcing process. Such an industry is often the most sophisticated buyer of the products and services it depends on. Its presence in a nation becomes important to developing competitive advantage in supplier industries. (1990, p. 149)

3 Data and methods

The hypothesis that we want to test is the following:

There is a positive relationship between the existence of advanced, domestic users and the competitiveness of domestic producers that supply these users with advanced equipment.

Most empirical work in this area is descriptive (case studies). These studies are often interesting and perceptive, but it is of course difficult to know how representative they are. To allow more general statements on the empirical relationships, we have, following Andersen *et al.* (1981), chosen a different method. The essence of this is the use of trade statistics to measure both the *competitiveness of the producers of technology and how advanced the domestic users are*. It is argued that one may approximate 'advanced domestic users' with 'internationally competitive (domestic) users'. This does not seem unreasonable. Firms that compete favourably on the world market, and want to continue to do so, have a clear incentive to acquire superior technology. This is also consistent with Porter's view (see the previous section).

A problem with this interpretation is that it limits the investigation to export products and home market sectors where the trade statistics allow a link to be made. Equipment that is used in many sectors, and sectors that mainly make use of such equipment, cannot be included. For instance, some well-known 'high-tech' industries, most notably the computer industry, had to be left out of the investigation for this reason. This, of course, does not mean that user–producer interaction is unimportant in these cases, just that these links cannot be explored by the methodology adopted here. Another consequence is that users in the service sectors of the economy (not covered by the trade statistics) are excluded. To remedy this somewhat, an attempt was made to construct special 'home-market indices' for three important service sectors. These sectors are health care, telecommunications and shipping (two of which are dominated by public-sector services).

Table 10.1 lists the 23 pairs of export products and home-market sectors. The sample includes 'advanced' export products for which the commodity classification (SITC, Revision 1, four-digit level) allowed a link to a home-market sector to be made.[5] The 23 pairs were divided into five groups, depending on *the character of the home market*. The first group includes export products with users in the food-producing sector, mostly agriculture. These user sectors are strongly regulated in all countries. As a consequence, there is in most cases little competition. The second and third groups include export products with users in the manufacturing sector, in 'traditional manufacturing' and 'transport equipment' respectively, for which the degree of competition is generally high. The services group is divided in two: shipping and public-sector services. Shipping is a typical global industry (Porter, 1990), with very competitive markets. Public-sector services, in contrast, have until recently been strongly regulated in all countries, with little competition domestically as well as internationally.[6]

To measure competitiveness, we use the familiar index for revealed comparative advantage (the RCA index, see Balassa, 1965). For a particular country and product, this index is the *ratio between the market share of the country on the world market for this particular product and the market share of the country on the world market for all products*. Letting X denote exports, i the exporting country and j the export product, the index for revealed comparative advantage (S) for country i in product j can be presented as follows:

$$S_{ij} = \frac{\dfrac{X_{ij}}{\sum_n X_{nj}}}{\dfrac{\sum_m X_{im}}{\sum_n \sum_m X_{nm}}}$$

where $n=1,\dots,i,\dots,N$, and $m=1,\dots,j,\dots,M$. This index has the property that the weighted mean is identical to unity for each country across all commodity groups, and for each commodity group across all countries. Thus, a country is said to have a revealed comparative advantage (be specialised) in a product if the RCA index exceeds unity.

It was argued above that one may approximate 'advanced domestic users'

Table 10.1 *Export products and home market sectors*

SITC (REV 1)	Export product	SITC (REV 1)	Home market sector
(A) Home market: agriculture			
6951	Hand tools for agriculture and forestry	04-08(-0814), 24	Agricultural products, wood products
7121, 7122	Agricultural machinery for preparing soil and harvesting	04-08(-0814)	Agricultural products
7123	Milking machines	02	Dairy products
7125	Tractors	04-08(-0814)	Agricultural products
7129	Agricultural machinery N.E.S.	04-08(-0814)	Agricultural products
7183	Food processing machinery	0(-00)	Food
7191	Heating and cooling equipment	01-03	Meat, dairy products, fish and eggs
(B) Home market: producers of traditional manufactures			
7151, 7152	Machine tools for working metals	69	Metal manufacturers
7171	Textile machinery	65	Textiles
7172	Leather machinery	61	Leather
7173	Sewing machinery	84	Clothing
7181	Paper working machinery	25, 64	Pulp and paper, paper products
7182	Printing machinery	829	Printed matter
7184, 7185	Construction and mining machinery, machinery for mineral crushing, etc.	27, 28	Crude minerals and metals
(C) Home market: producers of transport equipment			
6291	Rubber tyres and tubes	732-734	Road motor vehicles
7114	Aircraft engines	734	Aircraft
7115	Internal comb. engines	732	Road motor vehicles
7294	Automotive electrical equipment	732	Road motor vehicles

Table 10.1 (cont.)

SITC (REV 1)	Export product	SITC (REV 1)	Home market sector
(D) Home market: shipping			
735	Ships and boats		Shipping[a]
(E) Home market: (public sector) services			
54	Pharmaceuticals		Health[a]
7249	Telecommunications		Tele[a]
726	Electromedicals		Health[a]
8617	Medical instruments N.E.S.		Health[a]

Notes:

[a] For the definition of this indicator, see the text and appendix 1.

Explanatory note (example): 04-08 means 04, 05, 06 07, 08; 08(-0814) means 08 less 0814

with 'internationally competitive (domestic) users'. Thus, RCA indices were calculated also for the (k) home-market sectors ('home-market indices', S_{ik}). The indices for the service sectors were constructed to make them comparable to the RCA index. For instance, if the index for a specific country for shipping exceeds unity, this implies that the market share of the country for shipping services (merchant fleet registered in the country) exceeds the market share of the country for goods and services in general. For telecommunications and health services, which, until recently at least, were not traded on the world market to the same extent, the population of the country was used as a deflator. Thus, in these cases, a value larger than one implies that the per-capita 'quality' of these services in the country is higher than the OECD average. For telecommunications we used data for the number of telephone lines in the country, for health services we equated 'quality' with the economic resources devoted to this purpose. A problem with the latter may be that possible differences in health sector efficiency across the countries are not accounted for. (For details and sources, the reader is referred to the appendix.)

The trade data used in this chapter were collected from OECD Trade Series C, using the IKE Data Base at the University of Aalborg. Three years were included: 1965, 1973 and 1987. Since the theory is only expected to hold for developed countries, we excluded the industrially less developed of the OECD countries.[7] The countries included in the sample were: Canada, the USA, Japan, Austria, Belgium, Denmark, Finland, France, Germany, Italy, the Netherlands, Norway, Spain, Sweden, Switzerland and the UK.

4 Testing the hypothesis

In a general form, the model to be tested is the following

$$S_{ij}^t = f(S_{ik}^t, C_i^t), \quad dS_{ij}^t / dS_{ik}^t > 0 \tag{2}$$

This model includes two independent variables, the home-market index S_{ik} and a country-specific variable C_i. The inclusion of the latter reflects the possibility that there may exist additional, country-specific factors that affect comparative advantage, and which should be taken into account to avoid biased results.[8] For instance, a country with a comparative advantage in natural-resource based products (SITC 0–4), will by definition not have a comparative advantage for manufactured products (SITC 5–9). Since the dependent variable in all 23 cases belongs to the manufactured group, this implies that the dependent variable may be biased against countries spe-cialising in natural resource-based products. The inclusion of a country-specific constant term may correct for this type of bias.

In principle, the choice of functional form should be based on theory. But, as is common in testing of hypotheses, we have in this case no particular theoretical reasons for preferring one specific functional form. However, to get a better approximation to the assumption of normally distributed variables, a logarithmic form is preferred.[9] A Box–Cox test of the functional form came up with the same suggestion. Since there were zeros in the data matrix, we had to add a small positive number to all observations to allow the transformation to be made.

Thus, the tested model is as follows

$$\log(S_{ij}^t + 0.1) = c_i + a \log(S_{ik}^t + 0.1) \tag{3}$$

When in the following we refer to the variables S_{ij} and S_{ik}, it should be understood that these are in log-form, as in equation (3).

The questions we want to ask are:

1 Is there a positive relationship between the two specialisation indices, as argued by Linder and Porter, i.e., is the coefficient a positive?
2 Does the impact of the home-market variable (S_{ik}) decline over time, i.e., is the coefficient a smaller/less significant in 1987 than in earlier years?
3 To what extent does the introduction of a time-lag for the home-market variable (S_{ik}) improve the explanatory power of the model?
4 Are there significant differences across countries, or home markets, in the impact of the home-market variable (S_{ik})?

To answer the first two questions, equation (3) was tested on data for 1965, 1973 and 1987, with and without the country-specific variable C_i. The results are reported in table 10.2. In all cases, the coefficient a turned up significantly larger than zero at the 1 per cent level, as the Linder–Porter hypothesis would predict. The numerical estimate of a was remarkably stable across both time and differences in specification (the estimate varied between 0.43 and 0.49). Furthermore, there was no tendency towards a decrease in the numerical value of the estimate for a or its significance. The only notable difference between the tests reported in Table 10.2 relates to the impact of the country-specific variable C_i. In all cases the inclusion of this variable significantly increased the explanatory power of the model, but less so in 1987 than for earlier years, indicating that the importance of the country-specific factors may be reduced somewhat during this period.

Patterns of comparative advantage may be viewed as the result of a long-term historical process. Thus, there may be rather long lags present in the impact of user–producer interaction on comparative advantages. To shed some light on this issue we have included in Table 10.3 some tests where the independent variable is lagged one or two periods. Given that patterns of comparative advantage change only slowly, the home-market variables

Table 10.2 *The hypothesis tested*

1965		
$S_{ij} = -0.37 + 0.43\, S_{ik}$	$R^2 = 0.15\ (0.15)$	(2.1)
(7.54) (8.14)	SER$=0.90$	
* *		
$S_{ij} = C_i + 0.43\, S_{ik}$	$R^2 = 0.39\ (0.36)$	(2.2)
(9.00)	SER$=0.78$	
*		
1973		
$S_{ij} = -0.26 + 0.47\, S_{ik}$	$R^2 = 0.17\ (0.17)$	(2.3)
(5.69) (8.62)	SER $= 0.85$	
* *		
$S_{ij} = C_i + 0.49\, S_{ik}$	$R^2 = 0.36\ (0.33)$	(2.4)
(9.58)	SER $= 0.76$	
*		
1987		
$S_{ij} = -0.21 + 0.45\, S_{ik}$	$R^2 = 0.15\ (0.15)$	(2.5)
(4.76) (8.15)	SER $= 0.81$	
* *		
$S_{ij} = C_i + 0.49\, S_{ik}$	$R^2 = 0.33\ (0.30)$	(2.6)
(9.19)	SER $= 0.74$	
*		
$N = 368$		

Notes:
Method of estimation: Ordinary least squares, absolute *t*-values in
brackets. R^2 in brackets is adjusted for degrees of freedom.
* Significance at 1 per cent level.

should be expected to be strongly correlated across years, which was indeed
the case. To avoid multicollinearity in cases where two annual observations
of S_{ik} were to be included, we had to put one of them in first differences
(equations (3.2) and (3.4)).

The results of the tests with lagged variables in table 10.3 should be compared with the result without lags in table 10.2 (equation (2.6)). Then it
becomes clear that the explanatory power of the instantaneous relationship
(equation (2.6)) is not inferior to any of the lagged relationships, when
adjustments for differences in degrees of freedoms are made. Thus, surprisingly perhaps, there is not strong support in the data for long lags. This
is also confirmed by the low weights given to the lagged independent variables in equations (3.2) and (3.4).[10] These findings may indicate a two-way

Table 10.3 *Testing for lags (1987)*

$S_{ij87} = C_i + 0.39\, S_{ik65}$ (8.46) *	$R^2 = 0.31\ (0.28)$ (3.1) $SER = 0.75$
$S_{ij87} = C_i + 0.50\, S_{ik65} + 0.34\, (S_{ik87} - S_{ik65})$ (9.37) (3.85) * *	$R^2 = 0.34\ (0.30)$ (3.2) $SER = 0.74$
$S_{ij87} = C_i + 0.43\, S_{ik73}$ (8.64) *	$R^2 = 0.31\ (0.28)$ (3.3) $SER = 0.75$
$S_{ij87} = C_i + 0.49\, S_{ik73} + 0.39\, (S_{ik87} - S_{ik73})$ (9.20) (2.97) * *	$R^2 = 0.33\ (0.30)$ (3.4) $SER = 0.74$
$N = 368$	

Notes:
Method of estimation: Ordinary least squares, absolute t-values in brackets. R^2 in brackets is adjusted for degrees of freedom.
* Significance at 1 per cent level.

relationship between 'users' and 'producers', e.g., that the competitiveness of both parties is affected. As noted in section 2, this would not be inconsistent with the theory, but we will not discuss this issue further here. No attempt was made to test for the direction of causality.

In the tests reported so far we have implicitly assumed that all countries are identical except for the constant term, which was assumed to reflect sector-invariant, country-specific factors. Although we have no prior information that leads us to believe that the impact of the home-market variable on comparative advantage differs substantially across countries, this possibility cannot be excluded *a priori*. To account for this possibility we have included a test of the restriction that $a_1 = a_2 = \ldots = a_i = \ldots = a_N = a$. The results (table 10.4) are ambiguous. In no case can the hypothesis of a common coefficient a for all countries be rejected at the 1 per cent level. However, for 1965 and 1973 – but not for 1987 – the tests indicate that the hypothesis of a common coefficient can be rejected if the criterion of a 5 per cent significance level is adopted.

Table 10.5 lists the unrestricted estimates for the coefficient a. The results suggest that the countries of our sample may be divided roughly into four groups, depending on the strength of the relationship. For five countries (Japan, Denmark, Finland, Norway and Switzerland) the estimates of the coefficient a are positive, significant at the 1 per cent level, for all three years.

Table 10.4 *Testing for pooling*

	1965	1973	1987
16 country sample[a]	2.03 **	1.98 **	1.58
13 country sample[b]	1.53	1.44	1.59

Notes:
[a] All countries, F-statistics with degrees of freedom 15,336.
[b] All countries less Austria, France and UK, F-statistics
with degrees of freedom 12,273.
 * Significance at 1 per cent level.
** Significance at 5 per cent level.

Clearly, for these countries there is strong support for the hypothesis of a positive relationship between the two indices. Then follows a group of seven countries where there is some support, although weaker (positive, significant at the 10 per cent level, for at least two years): Canada, the USA, Germany, Italy, the Netherlands, Spain and Sweden. For Belgium and Austria too, a positive relationship was reported, though significant for one year only (at the 5 per cent and 10 per cent level, respectively). However, for France and the UK, the results give no support at all for the hypothesis of a positive relationship between the two indices. Taking this information into account, we repeated the test of the restriction (a common value of a for all countries) on a sample that excluded Austria, France and the UK. For this sample it was not possible to reject the restriction for any year (table 10.4).

The fact that the hypothesis is not empirically supported for some countries deserves an explanation. First, it cannot be excluded that this – to some extent at least – is the result of imperfect data or methods. It can be shown that there is a positive relationship between the statistical significance of the estimate a and the variance of the dependent variable, i.e., that countries with a 'flat' structure of export specialisation (low variance) generally have poor results. Low variance is a common problem in small samples, and it is possible that the results would have improved if the number 'pairs' included in the test had been larger. This was not possible with the available data. The problem of a 'flat' structure of export specialisation was especially pronounced for Belgium and France, and it is possible that the poor results for these countries may be explained by data limitation. This explanation is less probable for the UK and Austria, where the reported variances do not differ much from that of the sample as a whole. Unfortunately, we do not have a

Table 10.5 *Unrestricted estimates for $S_{ik}{}^a$*

	1965	1973	1987
Canada	0.28	0.20	0.42
	(1.52)	(1.01)	(2.15)
	***		**
USA	0.43	0.37	0.38
	(1.38)	(1.78)	(1.49)
	***	**	***
Japan	0.78	0.78	0.51
	(4.08)	(4.18)	(3.38)
	*	*	*
Austria	0.26	0.11	0.16
	(1.45)	(0.59)	(0.79)
	***		
Belgium	0.13	0.14	0.43
	(0.35)	(0.45)	(1.76)
			**
Denmark	0.58	0.56	0.77
	(3.95)	(3.72)	(4.78)
	*	*	*
Finland	0.50	0.47	0.51
	(3.69)	(2.93)	(2.78)
	*	*	*
France	−0.20	−0.09	−0.09
	(0.43)	(0.45)	(0.02)
Germany	0.31	0.42	0.31
	(1.34)	(1.50)	(0.80)
	***	***	
Italy	0.19	0.41	0.54
	(0.69)	(1.46)	(2.15)
		***	**
Netherlands	0.10	0.32	0.36
	(0.43)	(1.37)	(1.74)
		***	**
Norway	0.42	0.57	0.60
	(3.03)	(3.85)	(3.73)
	*	*	*
Spain	0.33	0.39	0.23
	(2.61)	(2.21)	(1.09)
	*	**	
Sweden	1.00	0.47	0.10
	(5.13)	(2.08)	(0.50)
	*	**	

Table 10.5 (*cont.*)

	1965	1973	1987
Switzerland	0.78	0.92	1.12
	(4.16)	(5.45)	(5.77)
	*	*	*
UK	−0.16	−0.17	0.40
	(0.66)	(0.60)	(0.78)

Notes:
[a] Estimated with country dummies (C_i).
Method of estimation: ordinary least squares, absolute
t-values in brackets.
* Significant at 1 per cent level, one-tailed test.
** Significant at 5 per cent level, one-tailed test.
*** Significant at 10 per cent level, one-tailed test.

good alternative explanation to offer. In general, the reported results show no clear relation to variables commonly used in cross-country analyses of specialisation patterns, such as – for instance – country size or income level (past or present). Arguably, a much more detailed analysis of economic, institutional and cultural factors seems to be required.[11]

The possibility of differences across home markets – in their impact on the competitiveness of suppliers – is perhaps more interesting. At least, here we have a well-argued case (Porter, 1990) for assuming that home markets exposed to international competition are more conducive than others in fostering internationally competitive suppliers. This can be tested in a similar way as for the differences across countries. The results (table 10.6) indicate that there may be significant differences between different types of home markets. Generally speaking, the relationship between the two indices appears to be much stronger for home-market sectors exposed to international competition (traditional manufacturing, production of transport equipment and shipping) than for the more 'sheltered' sectors (agriculture and public-sector services). For public-sector services the results may be affected by the problem of finding reliable indicators. Still, the results lend clear support to Porter's view on the importance of competition.

The results reported in this chapter may to some extent be compared with those reported by Anderson *et al.* (1981), although differences in both sample and methods exist.[12] In particular, it must be kept in mind that their sample was much smaller. Results for 1954, 1960, 1966 and 1972 were reported. In general, a significant relationship was found for approximately

Table 10.6 *Testing for differences between home market sectors, 1987[a]*

$$S_{ij} = C_i + 0.23 \ S_{ik}^{\text{agriculture}} + 0.52 \ S_{ik}^{\text{traditional}}$$
$$\phantom{S_{ij} = C_i + 0.23 } (2.74) \phantom{S_{ik}^{\text{agriculture}}} (5.74)$$
$$+ 0.68 \ S_{ik}^{\text{transport}} + 0.89 \ S_{ik}^{\text{traditional}}$$
$$ (6.59) \phantom{S_{ik}^{\text{transport}}} (5.12)$$
$$+ 0.55 \ S_{ik}^{\text{public}}$$
$$ (1.67)$$

$$R^2 = 0.36 \ (0.32)$$
$$\text{SER} = 0.72$$
$$N = 368$$
$$F_{(5,347)} = 3.89$$
$$(*)$$

Notes:
[a] Method of estimation: ordinary least squares, absolute t-values in brackets. R^2 in brackets is adjusted for degrees of freedom. $F_{(5,347)}$ is an F-test of whether there are significant differences across home market sectors.
* Significance at 1 per cent level.

half of the countries included in their sample. The countries for which they found no support for the hypothesis were Belgium, France and the United Kingdom (Austria was not included). This is in line with the results presented here. Furthermore, as in the present study, there was no sign of a weakening of the relationship: in fact, the best results were reported for the most recent years.

5 Concluding remarks

The view that the home market may have a positive impact on the competitiveness of domestic producers is by no means a new one. Indeed, it has been widely held for at least a century, although neo-classical trade theorists have condemned it as 'theoretically unsound'. Often it has been regarded as a pure cover for protectionism. More recently, however, Michael Porter (1990) has made a major effort to increase the credibility of this view, and with considerable success, especially among policy makers and industrialists.

This chapter has attempted to give an appraisal of the theoretical and empirical evidence on the hypothesis of a positive impact of the domestic market, through 'advanced domestic users', on the international competitiveness of a country (the 'Linder–Porter hypothesis'). It was suggested that a positive impact of this kind may be explained by a theory that focusses on *interaction between users and producers of technology* as a major impetus to technological change. Interaction, however, involves costs. It was argued that these are a decreasing function of both the stability of the user–producer relationship and the degree of 'proximity', defined to include factors such as language, the legal system, the education system, etc. Hence most stable user-producer relationships are of a *national* character. The above, together with the assumption that a country's comparative advantage in the long run will be in areas where its rates of learning and innovation are high (compared with other countries), suggests that countries in the long run tend to develop comparative advantages in areas where, by a comparative standard, there are many advanced domestic users.

Most previous empirical work in this area is of a descriptive character. This chapter, in contrast, has presented an econometric test of the hypothesis of a positive impact of advanced domestic users on competitiveness. The data set included 16 countries, 23 pairs of products and three selected years (1965, 1973 and 1987). The main empirical findings were:

1 There is strong support in the data for the hypothesis of a positive impact of advanced domestic users in competitiveness.
2 There is no evidence of a weakening of this relationship during the period 1965–87.
3 The time lag between the initial stimulus (from the domestic market) and the impact on competitiveness appears to be relatively short.
4 For most countries there is some support for the hypothesis. The most notable exceptions are France and the UK.
5 The relationship appears to be stronger in cases where the home market is exposed to international competition.

In general, these findings are consistent with the predictions made by Linder (1961) and Porter (1990).

The theoretical and empirical evidence presented in this chapter indicates that stable relationships between domestic users and producers of technology may have a positive impact on both technological progress and international competitiveness. This is especially so if these relationships develop in a competitive environment, i.e., that the positions of both users and producers may be contested. Thus, contrary to the belief of many economists, this approach does not favour protectionism. However, the emphasis in this approach on a competitive environment does not necessarily imply that every individual contract has to be open to public tender.

Arguably, a competition policy of this kind would in practice make stable user–producer relationships very difficult to maintain. Thus, the old Schumpeterian theme of the uneasy balance between static and dynamic efficiency may apply also in this case.

Appendix

The trade data used in this chapter were calculated from *OECD Trade Series C* (value data) using the IKE data base on trade statistics at the Aalborg University Centre. Data for health care were taken from *Health Care Systems in Transition*, OECD, Paris, 1990, data for merchant fleets and telephone lines were taken from *UN Statistical Yearbook*, various editions. Other data from *OECD National Accounts*.

Construction of home-market indicators

Telephones and health. T_j=telephone lines in country j ($i=1, \ldots, j, \ldots, n$); N_j=number of inhabitants in j. The index may then be written

$$I_j = \frac{T_j}{N_j} \bigg/ \frac{\sum_i T_i}{\sum_i N_i}$$

Similarly for health services, where T_j=total (public and private) expenses for health services in current prices in common currency.

Shipping. S_j=merchant fleet registered in country j, in 1000 tons; X_j=total exports of goods and services from country j in current prices in common currency

$$I_j = \frac{S_j}{\sum_i S_i} \bigg/ \frac{X_j}{\sum_i X_i}$$

Notes

This chapter builds up and extends earlier work by the author on the same subject (Fagerberg, 1992, 1994). The ideas owe much to discussions with members of the IKE group at the University of Aalborg, especially with Esben Sloth Andersen and Bent Dalum. I thank Bent Dalum and Vibeke Jakobsen, University of Aalborg, for assistance in data work. The final version benefited from comments from Daniele Archibugi. Financial support from the Nordic Economic Research Council is gratefully acknowledged.
1 Writers who have emphasised the importance of relations between firms or sectors, within the domestic economy, for industrialisation, growth and com-

petitiveness include Perroux (1955), Hirschman (1958), Lindar (1961) and Dahmén (1970). For an overview, see Dosi and Soete (1988).

2 Several factors may contribute to this. New knowledge may be very specific and difficult to transfer ('tacit knowledge'). Secrecy and legal procedures (patents and trade marks) are other means.

3 This way of looking at things shows some similarity with parts of the 'new-growth' literature (see Verspagen, 1992 for an overview). However, this literature (Romer, 1986, and others), as well as the older literature in this area (Arrow, 1962; Kaldor and Mirrlees, 1962), discusses externalities of *activities internal to firms* (investment and/or production). This chapter focuses on the effects of *interaction between different firms*.

4 Porter emphasises especially the importance of domestic rivalry, but acknowledges that, for small open economies, foreign competitors may serve a similar function (1990, p. 121).

5 'Advanced' is used here in a broad sense. Only products based on natural resources and relatively unsophisticated ('mature') manufactures were excluded.

6 As follows from table 10.1, we have assumed that the health sector is the 'user' of pharmaceutical products, although strictly speaking the final users are the individual patients/consumers. However, it is the health sector that decides on standards, etc., which is what matters in the present context.

7 Australia and New Zealand were excluded due to lack of data for some of the years covered by the investigation.

8 This is the so-called 'least-squares dummy variables method' (LSDV), which is developed for use in pooled data sets. For details, see Johnston (1984).

9 The index of revealed comparative advantage (S) has a skew distribution, with a long tail to the right. This creates problems in regression analysis, because it violates the assumption of normality. A logarithmic transformation of the data reduces this problem significantly.

10 In eqn. (3.2), the implicit weight is 0.16 for S_{ik65} and 0.34 for S_{ik67}, while in eqn. (3.4), the implicit weight is 0.10 for S_{ik73} and 0.39 for S_{ik87}.

11 See Tylecote (1993) for an interesting attempt to explain cross-country differences in the degree of interfirm collaboration with the help of some of these factors.

12 See Fagerberg (1992) for a more detailed presentation and discussion of the contribution by Andersen et al. (1981).

References

Abbott, T. A. (1991), 'Measuring high technology trade: contrasting international trade administration and Bureau of Census Methodologies and results', *Journal of Economic and Social Measurement*, 17: 17–44

Abbott, T. A., McGuchin, R., Herrick, P. and Norfolk, L. (1989), 'Measuring the trade balance in advanced technology products', Center for Economic Studies, Discussion Paper 89–1. Bureau of the Census, US Department of Commerce, Washington, DC

Abernathy, W. J. and Clark, K. B. (1985), 'Innovation: mapping the winds of creative destruction', *Research Policy*, 14: 3–22

Abernathy, W. and Hayes, R. (1980), 'Managing our way to economic decline', *Harvard Business Review*, July/August

Abramovitz, M. (1956), 'Resource and output trends in the United States since 1870', *American Economic Review, Papers and Proceedings*, 46(2): 5–23 (reprinted in Abramovitz, 1989)

(1986), 'Catching up, forging ahead and falling behind', *Journal of Economic History*, 46: 385–406

(1989), *Thinking About Growth*, Cambridge: Cambridge University Press

(1993), 'The search for the sources of growth: areas of ignorance, old and new', *Journal of Economic History*, 53(2): 217–43

Achilladelis, B. G., Schwarzkopf, A. and Lines, M. (1987), 'A study of innovation in the pesticide industry', *Research Policy*, 16(2): 175–212

(1990), 'The dynamics of technological innovation: the case of the chemical industry', *Research Policy*, 19(1): 1–35

Acs, Z. J. (1990), 'High technology networks in Maryland', paper presented at Montreal Conference on Network Innovators, May

Acs, Z. J. and Audretsch, D. B. (1988), 'Innovation in large and small firms: an empirical analysis', *American Economic Review*, 78(4): 678–90

Afuah, A. N. and Utterback, J. M. (1991), 'The emergence of a new supercomputer architecture', *Technological Forecasting and Social Change*, 40: 315–28

Aghion, P. and Howitt, P. (1993), 'A model of growth through creative destruction', in Foray, D. and Freeman, C. (eds.) *Technology and the Wealth of Nations*, London: Pinter

Alchian, A. (1950), 'Uncertainty, evolution and economic theory', *Journal of Political Economy*, 58: 211–22

Alderman, N. and Davies, S. (1990), 'Modelling regional patterns of innovation diffusion in the UK metal-working industries', *Regional Studies*, 24: 513–28

Alderman, N. and Wynarczyk, W. (1993), 'The performance of innovative small firms: a regional issue', in Swann, P. (ed.), *New Technology and the Firm*, London: Routledge

Allen, R. C. (1983), 'Collective invention', *Journal of Economic Behaviour and Organisation*, 4: 1–24

Amable, B. (1992), 'Radical and incremental innovation in a model of endogenous and unsteady Growth', paper at MERIT Conference, Maastricht, December

(1993a), 'National effects of learning, international specialisation and growth paths', in Foray, D. and Freeman, C. (eds.), *Technology and the Wealth of Nations*, London: Pinter

(1993b), 'Catch up and convergence: a model of cumulative growth', *International Review of Applied Economics*, 7: 1–25

Amable, B. and Boyer, R. (1992), 'L'Europe dans la competition technologique mondiale: quelques enjeux et propositions', CEPREMAP Working Paper no. 9202, Paris

Amendola, G., Dosi, G. and Papagni, E. (1993), 'The dynamics of international competitiveness', *Weltwirtschaftliches Archiv*, 129(3): 451–71

Amendola, G., Guerrieri, P. and Padoan, P.C. (1991), 'International patterns of technological accumulation and trade', *Journal of International and Comparative Economics*, 1(1): 173–97

Amendola G. and Perrucci, A. (1990), 'Specialisation and competitiveness of Italian industry in high-technology products: a new approach', Technology/Economy Programme (TEP), International Conference on 'Technology and Competitiveness', Paris, 25–27 June

(1992), 'European patterns of specialisation in high-technology products: a new approach', in: *Proceedings of the Joint EC-Leiden Conference on Science and Technology Indicators*, Leiden: DSWO Press

Amendola, M. and Gaffard, J. L. (1988), *The Innovation Choice: An Economic Analysis of the Dynamics of Technology*, Oxford: Blackwell

Ames, E. (1961), 'Research, invention, development and innovation', *American Economic Review*, 51(3): 370–81

Ames, E. and Rosenberg, N. (1963), Changing technological leadership and industrial growth', *Economic Journal*, 73: 13–31

Amin, M. and Goddard, J. B. (eds.) (1986), *Technological Change, Industrial Restructuring and Regional Development*, London: Allen and Unwin

Amsden, A. (1989), *Asia's Next Giant: South Korea and Late Industrialisation*, Oxford and New York: Oxford University Press

Andersen, E. S. (1991), 'Techno-economic paradigms as typical interfaces between producers and users', *Journal of Evolutionary Economics*, 1(2): 119–44

(1992a), 'Approaching national systems of innovation from the production and linkage structure', in Lundvall, B-Å. (ed.), *National Systems of Innovation*, London: Pinter

(1992b), 'The difficult jump from Walrasian to Schumpeterian analysis', paper at International Schumpeter Society Conference, Kyoto, August

(1993), *Schumpeter and the Elements of Evolutionary Economics*, London: Pinter

Andersen, E. S., Dalum, B. and Villumsen, G. (1981), *International Specialization and the Home Market*, Aalborg: Aalborg University Press

Angello, M. M. (1990), *Joseph Alois Schumpeter: A Reference Guide*, Berlin: Springer

Antonelli, C. (1986), 'The international diffusion of new information technologies', *Research Policy*, 15: 139–47

(1992), 'The economics of localised technological change: the evidence from information and communication technologies', University of Turin, Department of Economics

(1993), 'The dynamics of technological inter-relatedness: the case of information and communication technologies', in Foray, D. and Freeman, C. (eds.), *Technology and the Wealth of Nations*, London: Pinter

(1994), *The Economics of Localized Technological Change and Industrial Dynamics*, Boston: Kluwer Academic Publishers

Antonelli, C. (ed.) (1992), *The Economics of Information Networks*, Amsterdam: Elsevier

Antonelli, C. and Foray, D. (1991), 'The economics of intellectual property rights and systems of innovation', University of Turin, mimeo

Antonelli, C., Petit, P. and Tahar, G. (1992), *The Economics of Industrial Modernisation*, London: Academic Press

Aoki, M. (1986), 'Horizontal versus vertical information: structure of the firm', *American Economic Review*, 76(5): 971–83

(1988), *Information, Incentives and Bargaining in the Japanese Economy*, New York: Cambridge University Press

(1990), 'Towards an economic model of the Japanese firm', *Journal of Economic Literature*, 28: 1–27

(1991), 'The Japanese firm as a system: survey and research agenda', paper at Stockholm Conference on Japan, Stockholm School of Economics, September

Arcangeli, F., David, P. and Dosi, G. (eds.) (1986), *Frontiers on Innovation Diffusion*, Report of the Venice Conference on Innovation Diffusion, DAEST, Venice

Arcangeli, F., Dosi, G. and Moggi, M. (1991), 'Patterns of diffusion of electronics technologies: an international comparison', *Research Policy*, 20(6): 515–31

Archibugi, D. (1988a), 'In search of a useful measure of technological innovation', *Technological Forecasting and Social Change*, 34: 253–77

(1988b), 'The inter-industry distribution of technological capabilities. A case study in the application of the Italian patenting in the USA', *Technovation*, 7(3): 259–74

(1992), 'Patents as indicator of technological innovation', *Science and Public Policy*, 17: 357–68

Archibugi, D., Cesaratto, S. and Sirilli, G. (1987), 'Innovative activity, R&D and patenting: the evidence of the survey on innovation diffusion in Italy', *STI Review*, 2: 135–50

Archibugi, D. and Michie, J. (eds.) (1997), *Technology, Globalisation and Economic Performance*, Cambridge: Cambridge University Press

Archibugi, D. and Pianta, M. (1990), 'Patterns of national technological specialisation based on patent statistics', Conference on 'Global Research Strategy
and International Competitiveness', Reading, 9 January
Archibugi, D. and Pianta, M. (1991), 'Specialisation and size of technological activities in industrial countries: the analysis of patent data', Consiglio Nazionale
delle Ricerche. Istituto di Studi sulla ricerca e documentazione scientifica,
Rome
Archibugi, D. and Pianta, M. (1992), *The technological specialisation of advanced
countries, A report to the EC on International Science and Technology Activities,*
Dordrecht: Kluwer
(1997), 'Aggregate convergence and sectoral specialisation in innovation', this
volume
Archibugi, D. and Santarelli, E. (1989), 'Tecnologia e Struttura del Commercio
Internazionale: La Posizione dell'Italia', *Ricerche Economiche,* 43: 427–55
Arrighi, G. (1991), 'World income inequalities and the future of socialism', *New
Left Review,* 31: 30–65
Arrow, K. (1962), 'The economic implications of learning by doing', *Review of
Economic Studies,* 29: 155–73
(1994), 'The production and distribution of knowledge', in Silverberg, G. and
Soete, L. (eds.), *The Economics of Growth and Technical Change: Technologies,
Nations, Agents,* Aldershot: Edward Elgar
Arthur, W. B. (1983), 'Competing techniques and lock-in by historical events. The
dynamics of allocation under increasing returns', IIASA, Luxemburg; rev.
edn. CEPR, Stanford University 1985
(1986), 'Industry, location and the importance of history', Center for Economic
Policy Research Paper 84, Stanford University
(1988), 'Competing technologies: an overview', in Dosi, G. *et al.* (eds.), *Technical
Change and Economic Theory,* London: Pinter
(1989), 'Competing technologies, increasing returns and lock-in by historical
events', *Economic Journal,* 99(1): 116–31
Arthur, W. B; Ermoliev, Y. M. and Kaniovski, Y. M. (1987), 'Path-dependence
processes and the emergence of macro-structure', *European Journal of
Operational Research,* 30: 294–303
Arundel, A. and Soete, L. (eds.) (1993), 'An integrated approach to European inno-
vation and technology diffusion policy: a Maastricht memorandum',
University of Limburg: MERIT
Atkinson, A. and Stiglitz, J. (1969), 'A new view of technological change', *Economic
Journal,* 78: 573–8
Auzeby, F. and François, J-P. (1992), 'Technological innovation in the French indus-
try', *STI Review,* 11: 118–24
Ayres, R. V. (1991a), 'Information, computers, CIM and productivity', in OECD
(1991b), *Technology and Productivity,* Paris: OECD
(1991b), *Computer Integrated Manufacturing,* vol. I, London: Chapman and
Hall/IIASA (4 vols.)
Baba, Y. (1985), 'Japanese colour TV firms. Decision-making from the 1950s to the
1980s', D.Phil. dissertation, Brighton: University of Sussex
Baba, Y. and Takai, S. (1990), 'Information technology introduction in the big

banks: the case of Japan', in Freeman, C. and Soete, L. (eds.), *New Explorations in the Economics of Technological Change*, London: Pinter

Bailey, M. W. and Chakrabarti, A. K. (1985), 'Innovation and productivity in US industry', *Brookings Papers on Economic Activity*, 2: 609–32

Balassa, B. (1965), 'Trade liberalisation and revealed comparative advantage', *The Manchester School of Economic and Social Studies*, 33(2): 99–124

Barras, R. (1986), 'Towards a theory of innovation in services', *Research Policy*, 15(4): 161–73

(1990), 'Interactive innovation in financial and business services: the vanguard of the service revolution', *Research Policy*, 19(3): 215–37

Barro, R.J. (1991), 'Economic growth in a cross section of countries', *Quarterly Journal of Economics*, 105: 407–43

Basberg, B. (1987), 'Patents and the measurement of technological change: a survey of the literature', *Research Policy*, 12(2–4): 131–43

Baumol, W.J., Blackman, B.S.A. and Wolff, E.N. (1989), *Productivity and American Leadership: the Long View*, Cambridge, Mass.: MIT Press

Beelen, E. and Verspagen, B. (1994), The role of convergence in trade and sectoral growth', in Fagerberg, J., Verspagen, B. and von Tunzelmann, N. (eds.), *The Dynamics of Technology, Trade and Growth*, Aldershot: Edward Elgar

Bell, M. (1984), 'Learning and accumulation of industrial and technological capability in developing countries', in King, K. and Fransman, M. (eds.) *Technological Capacity in the Third World*, London: Macmillan

(1991), 'Science and technology policy research in the 1990s: key issues for developing countries', Brighton: University of Sussex, SPRU

Bell, M. and Pavitt, K. (1992), 'National capacities for technological accumulation', Washington DC: World Bank Conference on Development Economics

(1993), 'Technological accumulation and industrial growth: contrasts between developed and developing countries', *Industrial and Corporate Change*, 2: 157–210

Bell, R. M. N., Cooper, C. M., Kaplinsky, R. M. and Wit Sakyarakwit (1976), *Industrial Technology and Employment Opportunity: A Study of Technical Alternatives for Can Manufacture in Developing Countries*, Geneva: International Labour Organisation

Benhabib, J. and Spiegel, M.M. (1992), 'The role of human capital and political instability in economic development', New York University, C.V. Starr Center for Applied Economics, Research paper 92–24

Bernal, J. D. (1939), *The Social Function of Science*, London: Routledge and Kegan Paul

(1970), *Science and Industry in the Nineteenth Century*, Bloomington: Indiana University Press, 3rd edn

Bertin, G. and Wyatt, S. (1988), *Multinationals and Industrial Property*, Hemel Hempstead: Harvester-Wheatsheaf

Bessant, J. (1991), *Managing Advanced Manufacturing Technology*, Oxford: Blackwell

Bessant, J. and Haywood, W. (1991), 'Mechatronics and the machinery industry', in Freeman, C. .et al. (eds.) *Technology and the Future of Europe: Global Competition and the Environment in the 1990s*, London and New York: Pinter

Bessant, J., Burnell, J., Hardy, R. and Webb, S. (1993), 'Continuous innovation in British manufacturing', *Technovation*, 13(4): 241–54

Bianchi, P. and Bellini, N. (1991), 'Public policies for local networks of innovators', *Research Policy*, 20(5): 487–97

Bijker, W. E. and Law, J. (eds.) (1992), *Shaping Technology, Building Society*, Cambridge, Mass.: MIT Press

Blaug, M. (1978), *Economic Theory in Retrospect*, Cambridge University Press

Blomstrom, M. and Wolff, E.N. (1994), 'Multinational corporations and productivity convergence in Mexico', in Baumol, W.J., Nelson, R.R. and Wolff E.N. (eds.), *Convergence of Productivity: Cross-National Studies and Historical Evidence*, New York: Oxford University Press

Boyer, R. (1988), 'Technical change and the theory of regulation', in Dosi, G. *et al.* (eds.), *Technical Change and Economic Theory*, London: Pinter; New York: Columbia University Press

——— (1993), 'The models revolution', Introduction to part II, in Foray, D. and Freeman, C. (eds.), *Technology and the Wealth of Nations*, London: Pinter

Boyer, R. and Petit, P. (1989), 'Kaldor's growth theories: past, present and prospects', in Semmler, W. and Neil, E. (eds.), *Nicholas Kaldor and Mainstream Economics*, London: Routledge

Bradford, C. and Branson,W.H. (1987), *Trade and Structural Change in Pacific Asia*, Chicago: National Bureau of Economic Research and University of Chicago Press

Brady, T. M. (1986), 'New Technology and Skills in British Industry', Skills Series 5, Manpower Services Commission

Brady, T. M. and Quintas, P. (1991), 'Computer software: the IT constraint', in Freeman, C. *et al.* (eds.), *Technology and the Future of Europe*, London: Pinter

Braun, E. and MacDonald, S. (1978), *Revolution in Miniature: The History and Impact of Semi-Conductor Electronics*, Cambridge: Cambridge University Press

Breschi, S. and Mancusi, M.L. (1996), 'Il modello di specializzazione tecnologica dell'Italia', *Quaderni di politica industriale*, 8, Mediocredito Centrale, Rome

Bressand, A. (1990), 'Electronic cartels in the making?', *Transatlantic Perspectives*, 21: 3–6

Bressand, A. and Kalypso, N. (eds.) (1989), *Strategic Trends in Services: An Inquiry into the Global Service Economy*, New York: Harper and Row

Buigues, P., Ilzkovitz, F. and Lebrun, J.-F. (1990), Industrieller Strukturwandel im europäischen Binnenmarkt; Anpassungsbedarf in den Mitgliedstaaten, in: Europäische Wirtschaft – Soziales Europa. Commission of the European Communities, Brussels, Luxemburg

Burns, T. and Stalker, G. M. (1961), *The Management of Innovation*, London: Tavistock

C. I. News (1993), Bulletin of the CIRCA, Continuous Improvement Network 2

Callon, M. (1993), 'Variety and irreversibility in networks of technique conception and adoption', Foray, D. and Freeman, C. (eds.), *Technology and the Wealth of Nations*, London: Pinter

Camagni, R. (ed.) (1991), *Innovation Networks: Spatial Perspectives*, London: Belhaven Press

Camagni, R. *et al.* (1984), *Il Robot Italiano. Produzione e Mercato della Robotica Industriale*, Milano: Il sole 24 Ore

Cantwell J. (1989), *Technological Innovation and Multinational Corporations*, Oxford: Basil Blackwell

(1991a), 'The international agglomeration of R&D', in Casson, M. (ed.), *Global Research Strategy and International Competitiveness*, Oxford: Blackwell

(1991b), 'Historical trends in international patterns of technological innovation', in Foreman-Peck, J. (ed.), *New Perspectives on the Late Victorian Economy: Essays in Quantitative Economic History 1860–1914*, Cambridge: Cambridge University Press

(1992), 'The internationalisation of technological activity and its implications for competitiveness', in Grandstrand, O., Hakanson, L. and Sjolander, S. (eds.), *Technology Management and International Business*, Chichester: Wiley

(ed.) (1993), *Transnational Corporations and Innovatory Activities*, London: Routledge

Carlsson, B. and Jacobsson, S. (1993), 'Technological systems and economic performance: the diffusion of factory automation in Sweden', in Foray, D. and Freeman, C. (eds.), *Technology and the Wealth of Nations*, London: Pinter, chapter 4

Carlsson, B. and Stankiewicz, R. (1991), 'On the nature, formation and composition of technological systems', *Journal of Evolutionary Economics*, 1(2): 93–119

Carter, C. F. and Williams, B. R. (1957), *Industry and Technical Progress*, Oxford: Oxford University Press

(1958), *Investment in Innovation*, Oxford: Oxford University Press

(1959a), *Science and Industry*, London: Oxford University Press

(1959b), 'The characteristics of technically progressive firms', *Journal of Industrial Economics*, 7(2): 87–104

Cassiolato, J. (1992), 'The user–producer connection in hi-tech: a case study of banking automation in Brazil', in Schmitz, H. and Cassiolato, J., *High-Tech for Industrial Development*, London: Routledge

Casson, M. (ed.) (1991), *Global Research Strategy and International Competitiveness*, Oxford: Blackwell

Cawson, A. and Holmes, P. (1991), 'The new consumer electronics', in: Freeman, C., Sharp, M. and Walker, W. (eds.), *Technology and the Future of Europe: Global Competition and the Environment in the 1990s*, London and New York: Pinter

CEPII (1983), *Economie mondiale: la montée des tensions*, Paris: Economica

(1989), *Commerce international: la fin des avantages acquis,* Paris: Economica

Cesaratto, S. and Sirilli, G. (1992), 'Some results of the Italian survey on technological innovation', *STI Review*, 11: 80–95

Chandler, A. D. (1977), *The Invisible Hand: The Managerial Revolution in American Business*, Cambridge, Mass.: Belknap Press, Harvard University

(1990), *Scale and Scope: The Dynamics of Industrial Capitalism*, Cambridge, Mass.: Belknap Press, Harvard University

(1992), 'Corporate strategy, structure and control methods in the United States in the 20th century', *Industrial and Corporate Change*, 1: 263–84

Chenery, H., Robinson, S., and Syrquin, M. (1986), *Industrialisation and Growth: A Comparative Study*, New York: Oxford University Press

Chesnais, F., (1986), 'Science, technology and competitiveness', *STI Review*, 1: 97–148

—— (1988a), Multinational Enterprises and the International Diffusion of Technology, in Dosi, G. *et al.* (eds.), *Technical Change and Economic Theory*, London: Pinter

—— (1988b), 'Technical cooperation agreements between firms', *STI Review*, 4: 57–119

—— (1992), 'National systems of innovation, FDI and the operations of MNEs', in Lundvall, B-Å. (ed.), *National Systems of Innovation*, London: Pinter

Chiaromonte, F. and Dosi, G. (1993), 'The micro foundations of competitiveness and their macroeconomic implications', in Foray, D. and Freeman, C. (eds.), *Technology and the Wealth of Nations*, London: Pinter, chapter 5

Chow, P. and Kellman, H.M. (1993), *Trade: the Engine of Growth in East Asia*, Oxford: Oxford University Press

Christensen, J. L. (1992), 'The role of finance in national systems of innovation', in Lundvall, B-Å. (ed.), *National Systems of Innovation*, London: Pinter

Clark, J. A., Freeman, C. and Soete, L. L. G. (1981), 'Long waves, inventions and innovations', *Futures*, 13(4): 308–22

Clark, K., Fujimoto, T. and Chew, W. (1987), 'Product development in the world auto industry', *Brookings Papers on Economic Activity*, no. 3

Clark, K. B. and Fujimoto, T. (1989), 'Lead time in automobile product development: explaining the Japanese advantage', *Journal of Engineering and Technology Management*, 6: 25–58

Clark, N. (1990), 'Evolution, complex systems and technological change', *Review of Political Economy*, 2(1): 26–42

Coase, R. H. (1937), 'The nature of the firm', *Economica*, 16: 386–405

—— (1988), 'The nature of the firm: origin', *Journal of Law, Economics and Organisation*, 4(1): 3–47

Coe, D.T. and Helpman, E. (1993), 'International R&D spillovers', Cambridge, Mass.: NBER Working Paper No. 4444, August

Coe, D.T., Helpman, E. and Hoffmaister, A.W. (1995), 'North–South R&D spillovers', Cambridge, Mass.: NBER Working Paper No. 5048, March

Cohen, W.M. and Levinthal, D.A. (1989), 'Innovation and learning: the two faces of R&D', *Economic Journal*, 99: 569–96

Cohendet, P. M. *et al.* (1987), *Les Matériaux Nouveaux*, Paris: Economica

Cohendet, P., Héraud, J.-A. and Zuscovitch, E. (1993), 'Technological learning, economic networks and innovation appropriability', in Foray, D. and Freeman, C. (eds.), *Technology and the Wealth of Nations*, London: Pinter

Commission of the EC (1988), 'The economics of 1992', *European Economy*, 35 (March)

Coombs, R. and Richards, A. (1991), 'Technologies, products and firms' strategies: Part 1 – a framework for analysis', *Technology Analysis and Strategic Management* 13(1): 77–86; Part 2 – analysis of three cases', *Technology Analysis and Strategic Management*, 3(2): 157–75

Coombs, R., Saviotti, P. and Walsh, V. (1987), *Economics and Technological Change*, London: Macmillan
(1992), *Technological Change and Company Strategies*, London: Harcourt Brace
Cooper, C. (ed.) (1973), *Science, Technology and Development*, London: Frank Cass
(1974), 'Science policy and technological change in undeveloped economies', *World Development*, 2(3)
Cooper, C. M. and Sercovitch, F. (1971), *The Channels and Mechanisms for the Transfer of Technology from Developed to Developing Countries*, Geneva: UNCTAD
Corbett, J. and Mayer, C. (1991), 'Financial Reform in Eastern Europe', Discussion Paper No. 603, Centre for Economic Policy Research (CEPR), London
Cowan, R. (1990), 'Nuclear power reactors: a study in technological lock-in', *Journal of Economic History*, 50: 541–67
Cressey, P. and Williams, R. (1990), *Participation in Change: New Technology and the Role of Employee Involvement*, Dublin: European Formulation for the Improvement of Living and Working Conditions
Cusumano, M. A. (1991), *Japan's Software Factories: A Challenge to US Management*, Oxford: Oxford University Press
Cyert, R. M. and March, J. G. (1963), *A Behavioral Theory of the Firm*, Englewood Cliffs, New Jersey: Prentice-Hall
Dahlman, C., Ross-Larsen, B. and Westphal, L. (1987), 'Managing technological development: lessons from newly industrialising countries', *World Development*, 15: 759–775
Dahmén, E. (1950), *Entrepreneurial Activity in Swedish Industry 1909–1939*, Stockholm: IUL
Dahmén, E. (1970), *Entrepreneurial Activity and the Development of Swedish Industry 1919–1939*, Homewood, American Economic Association Translation Series
(1988), 'Development blocks in industrial economics', *Scandinavian Economic History Review*, 1: 3–14
Dalum, B. (1992), 'Export specialisation, structural competitiveness and national systems of innovation', in Lundvall, B.-Å. (ed.), *National Systems of Innovation: Towards a Theory of Innovation and Interactive Learning*, London: Pinter
Daniels, P. (1993), 'Research and development, human capital and trade performance in technology-intensive manufacturers: a cross-country analysis', *Research Policy*, 22: 207–41
Dankbaar, B. (1993), *Economic Crisis and Institutional Change: The Crisis of Fordism from the Perspective of the Automobile Industry*, Maastricht: University Press
Darwin, C. (1859), *The Origin of Species*, London: Murray
Dasgupta, P. and David, P.A. (1987), 'Information disclosure and the economics of science and technology', in Feiwel, G. (ed.), *Arrow and the Ascent of Modern Economic Theory*, New York: New York University Press
Dasgupta, P. and Stoneman, P. (eds.) (1987), *Economic Policy and Technological Progress*, Cambridge University Press

David, P. A. (1976), *Technical Choice, Innovation and Economic Growth*, Cambridge University Press

(1985), 'Clio and the economics of QWERTY', *American Economic Review*, 75(2): 332–7. (An extended version is published in Parker, W. N. (ed.) (1986), *Economic History and the Modern Economist*, Oxford: Blackwell)

(1986a), 'Narrow windows, blind giants and angry orphans: the dynamics of systems rivalries and dilemmas of technology policy', CEPR Working Paper 10, Stanford University

(1986b), 'Technology diffusion, public policy and industrial competitiveness', in Landau, R. and Rosenberg N. (eds.), *The Positive Sum Strategy*, Washington, DC: National Academy of Sciences

(1991), 'Computer and dynamo: the modern productivity paradox in a not-too-distant mirror', in *Technology and Productivity: The Challenges for Economic Policy*, Paris: OECD

(1992), 'Knowledge, property and the system dynamics of technological change', paper presented to the World Bank Conference on Development Economics, Washington, DC, 30 April–1 May

(1993), 'Path-dependence and predictability in dynamic systems with local network externalities: a paradigm for historical economics', in Foray, D. and Freeman, C. (eds.), *Technology and the Wealth of Nations*, London: Pinter

David, P. and Greenstein, S. (1990), 'The economics of compatibility standards: an introduction to recent research', *Economics of Innovation and New Technology*, 1(1): 3–43

David, P. and Steinmuller, E. (1990), 'The ISDN bandwagon is coming, but who will be there to climb aboard? Quandaries in the economics of data communication networks', *Economics of Innovation and New Technology*, 1(1): 43–63

Davies, S. (1979), *The Diffusion of Process Innovations*, Cambridge: Cambridge University Press

DeBresson, C. (1989), 'Breeding innovation clusters: a source of dynamic development', *World Development*, 17(1): 1–6

(1993), *Comprehendre Le Changement Technique*, Ottowa: Les Presses de L'Université d'Ottowa

DeBresson, C. and Amesse, F. (1991), 'Networks of innovators: a review and introduction to the issue', *Research Policy*, 20(5): 363–79

De la Mothe, J. (ed.) (1990), *Science, Technology and Free Trade*, London: Pinter

De Long, J. B. and Summers, L. H. (1991), 'Equipment investment and economic growth', *Quarterly Journal of Economics*, 106(2): 445–502

Deiaco, E. (1992), 'New views on innovative activity and technological performance: the Swedish innovation survey', *STI Review*, 11: 36–62

Denison, E. F. (1967), *Why Growth Rates Differ: Post-War Experience in Nine Western Countries*, Washington, DC: Brookings Institution

Dertouzos, M., Lester, R. and Solow, R. (eds) (1989), 'Made in America', Report of the MIT Commission on Industrial Productivity, Cambridge, Mass.: MIT Press

Deutsche Bundesbank (1994), 'Reale Wechselkurse als Indikatoren der internationalen Wettbewerbsfähigkeit', *Monatsbericht*: 47–60

Dockès, P. (1991), 'Histoire "raisonnée" et économie historique', *Revue Économique*, 2: 181–208

Dodgson, M. (1991), *The Management of Technological Learning: Lessons from a Biotechnology Company*, Berlin: De Gruyter

    (1993), *Technological Collaboration in Industry*, London: Routledge

Dodgson, M. (ed.) (1989), *Technology Strategy and the Firm: Management and Public Policy*, London: Longman

Dodgson, M. and Rothwell, R. (eds.) (1994), *The Handbook of Industrial Innovation*, Aldershot: Edward Elgar

Dodgson, M. and Sako, M. (1993), 'Learning and trust in inter-firm linkages', paper at Conference on 'Technological Collaboration: Networks, Institutions and States', University of Manchester, April

Dollar, D. and Wolff, E.N. (1988), 'Convergence of industry labor productivity among advanced economies, 1963–82', *Review of Economics and Statistics*, 70: 549–58

    (1993), *Competitiveness, Convergence and International Specialisation*, Cambridge, Mass., MIT Press

Dore, R. (1973), *British Factory – Japanese Factory: The Origins of National Diversity in Industrial Relations*, Berkeley: University of California Press

    (1985), 'The sources of the will to innovate', Papers in Science and Technology Policy, 4, London: Imperial College

    (1987), *Taking Japan Seriously: A Confucian Perspective on Leading Economic Issues*, London: Athlone Press

Dosi, G. (1982), 'Technological paradigms and technological trajectories: a suggested interpretation of the determinants and directions of technical change', *Research Policy*, 11 (3, June): 147–62

    (1984), *Technical Change and Industrial Transformation*, London: Macmillan

    (1988), 'Sources, procedures and microeconomic effects of innovation', *Journal of Economic Literature*, 36: 1126–71

    (1991), 'The research in innovation diffusion: an assessment', in Nakicenovic, N. and Grübler, A. (eds.), *Diffusion of Technologies and Social Behaviour*, Berlin: Springer

Dosi, G. and Egidi, M. (1991), 'Substantive and procedural uncertainty: an exploration of economic behaviour in changing environments', *Journal of Evolutionary Economics*, 1(2): 145–68

Dosi, G. and Fabiani, S. (1994), 'Convergence and divergence in the long-term growth of open economies', in Silverberg, G. and Soete, L. (eds.), *The Economics of Growth and Technical Change. Technologies, Nations, Agents*, Aldershot: Edward Elgar

Dosi, G. and Freeman, C. (1992), 'The diversity of development patterns: on the processes of catching up, forging ahead and falling behind', paper prepared for the International Economics Association meeting, Varenna, 1–3 October

Dosi, G., Freeman, C., Nelson, R., Silverberg, G. and Soete, L. (eds.) (1988), *Technical Change and Economic Theory*, London: Pinter; New York: Columbia University Press

Dosi, G., Giannetti, G. and Toninelli, P. A. (eds.) (1992), *Technology and Enterprise in a Historical Perspective*, Oxford: Oxford University Press

Dosi, G. and Orsenigo, L. (1988), 'Coordination and transformation: an overview of structure, behaviour and change in evolutionary environments', in Dosi *et al.*, (eds.) *Technical Change and Economic Theory*, London: Pinter; New York: Columbia University Press

Dosi, G., Pavitt K., and Soete L. (1990), *The Economics of Technical Change and International Trade*, Hemel Hempstead: Harvester Wheatsheaf

Dosi, G., Silverberg, G. and Orsenigo, L. (1988), 'Innovation, diversity and diffusion: a self-organisation model', *Economic Journal*, 98: 1032–54

Dosi, G. and Soete, L. (1988), 'Technical change and international trade', in: Dosi, G., *et al.* (eds.) *Technical Change and Economic Theory*, London: Pinter; New York: Columbia University Press

Drabek, Z. and Greenaway, D. (1984), 'Economic integration and infra-industry trade: the CMEA and EEC compared', *Kyklos*, 37: 444–69

Dunning, J. H. (1988), *Multinationals, Technology and Competitiveness*, London: Allen and Unwin

Eaton, J. and Kortum, S. (1995), 'Engines of growth: domestic and foreign sources of innovation', Boston, Boston University mimeo

Edquist, C. (1989), 'The realm of freedom in modern times: new technology in theory and practice', Tema T, Report 18, University of Linkøping, Department of Technology and Social Change

Eliasson, G. (1986), 'Micro-heterogeneity of firms and stability of industrial growth', in Day, R. and Eliasson, G. (eds.), *The Dynamics of Market Economies*, Amsterdam: North Holland

(1988), 'Schumpeterian innovation, market structure, and the stability of industrial development', in Hanusch, H. (ed.), *Evolutionary Economics: Applications of Schumpeter's Ideas*, Cambridge: Cambridge University Press, 151–99

(1990), 'The firm as a competent team', *Journal of Economic Behavior and Organisation*, 13(3)

(1991a), 'Deregulation, innovative entry and structural diversity as a source of stable and rapid economic growth', *Journal of Evolutionary Economics*, 1(1): 49–63

(1991b), 'Modelling economic change and restructuring', in P. de Wolf (ed.), *Competition in Europe*, Kluwer: Dordrecht

(1991c), 'Modelling the experimentally organised economy: complex dynamics in an empirical micro–macro model of endogenous economic growth', *Journal of Economic Behaviour and Organisation*, 16: 153–82

(1992), 'Business competence, organisational learning and economic growth: establishing the Smith–Schumpeter–Wicksell (SSW) connection', in Scherer, F. and Perlman, M. (eds.), *Entrepreneurship, Technological Innovation and Economic Growth*, Ann Arbor: University of Michigan Press

Enos, J. L. (1962), *Petroleum Progress and Profits: A History of Process Innovation*, Cambridge, Mass.: MIT Press

Ergas, H. (1984), 'Why do some countries innovate more than others?', Center for European Policy Studies Paper 5

(1987), 'The importance of technology policy', in Dasgupta, P. and Stoneman, P. (eds.), *Economic Policy and Technological Performance*, Cambridge: Cambridge University Press

Ernst, D. and O' Connor, D. (1992), *Competing in the Electronics Industry*, Paris, OECD

Evenson, R. E. (1981), 'Benefits and obstacles to appropriate agricultural technology', *The Annals of the American Academy of Political and Social Science*: 54–67

(1993), 'Patents, R&D and invention potential: international evidence,' *American Economic Review*, 83: 463–8

Fagerberg, J. (1987), 'A technology gap approach to why growth rates differ', *Research Policy*, 16(2–4): 87–101

(1988a), 'International competitiveness', *Economic Journal*, 98: 355–74

(1988b), 'Why growth rates differ', in Dosi, G. *et al.* (eds.), *Technical Change and Economic Theory*, London: Pinter

(1992), 'The home market hypothesis re-examined: the impact of domestic-user-producer interaction in exports', in Lundvall, B-Å. (ed.), *National Systems of Innovation*, London: Pinter

(1994), 'Technology and international differences in growth rates', *Journal of Economic Literature*, 32: 1147–75

Fagerberg, J., Verspagen, B. and von Tunzelmann, N. (eds.) (1994), *The Dynamics of Technology, Trade and Growth*, Aldershot: Edward Elgar

Faulkner, W. (1986), 'Linkage between academic and industrial research: the case of biotechnological research in the pharmaceutical industry', D.Phil. thesis, Brighton: University of Sussex

Fecher, F. and Perelman, S. (1989), 'Productivity growth, technological progress and R&D in OECD industrial activities', in *Public Finance and Steady Economic Growth*, The Hague: International Institute of Public Finance

Feller, I., Madden, P., Kaltreider, L., Moore, D. and Sims, L. (1987), 'The new agricultural research and technology transfer policy agenda', *Research Policy*, 16(6): 315–27

Flamm, K. (1987), *Targeting Technology, National Policy and International Competition in Computers*, Washington, DC: Brookings Institute

(1988), *Creating the Computer: Government, Industry and High Technology*, Washington, DC: Brookings Institute

Fleck, J. (1983), 'Robots in manufacturing organisations', in Winch, G. (ed.), *Information Technology in Manufacturing Processes*, London: Rossendale

(1988), 'Innofusion or diffusation? The nature of technological development in robotics', ESRC Programme on Information and Communication Technologies (PICT), Working Paper series, University of Edinburgh

(1993), 'Configurations crystallising contingency', *International Journal of Human Factors in Manufacturing*, 3(1): 15–36

Foray, D. (1987), *Innovation Technologique et Dynamique Industrielle*, Lyon: Presses Universitaires de Lyon

(1991), 'The secrets of industry are in the air: industrial cooperation and the organisational dynamics of the innovative firm', *Research Policy*, 20(5): 393–405

(1992), 'The economics of intellectual property rights and systems of innovation: the inevitable diversity', paper at MERIT Conference, Maastricht, December

(1993), 'General introduction', in Foray, D. and Freeman, C. (eds.), *Technology and the Wealth of Nations*, London: Pinter

Foray, D. and Freeman, C. (eds.) (1993), *Technology and the Wealth of Nations*, London: Pinter

Foray, D. and Grübler, A. (1990), 'Morphological analysis: diffusion and lock-out of technologies: ferrous casting in France and the FRG', *Research Policy*, 19(6): 535–50

Frame, J. (1991), 'Modelling national technological capacity', *Scientometrics*, 22(3): 327–39

Franko, L. (1989), 'Global corporate competition: who's winning, who's losing, and the R&D factor as one reason why', *Strategic Management Journal*, 10: 449–74

Fransman, M. (1990), *The Market and Beyond: Cooperation and Competition in IT in the Japanese System*, Cambridge University Press

Freeman, C. (1962), 'Research and development: a comparison between British and American industry', *National Institute Economic Review*, 20: 21–39

(1971), 'The role of small firms in innovation in the UK since 1945', Bolton Committee Research Report 6, London: HMSO

(1974), *The Economics of Industrial Innovation*, first edn., Harmondsworth: Penguin; second edn., London: Pinter (1982)

(1987), *Technology Policy and Economic Performance: Lessons from Japan*, London: Pinter

(1991a), 'Networks of innovators: a synthesis of research issues', *Research Policy*, 20(5): 499–514

(1991b), 'Innovation, changes of techno-economic paradigm and biological analogies in economics', *Revue Économique*, 42(2): 211–32

(1992), *The Economics of Hope*, London: Pinter

(1993), 'The political economy of the long wave', European Association of Political Economy, Barcelona Conference, October

(1994), 'Innovation and growth', in Dodgson, M. and Rothwell, R. (eds.), *The Handbook of Industrial Innovation*, Aldershot: Edward Elgar

(1995), 'The national system of innovation in historical perspective', *Cambridge Journal of Economics*, 19(1): 5–24

Freeman, C. (ed.) (1984), *Long Waves in the World Economy*, London: Pinter

(1987), *Output Measurement in Science and Technology*, Amsterdam: North-Holland

Freeman, C., Clark, J. and Soete, L. (1982), *Unemployment and Technical Innovation*, London: Pinter

Freeman, C., Fuller, J. K. and Young, A. J. (1963), 'The plastics industry: a comparative study of research and innovation', *National Institute Economic Review*, 26: 22–62

Freeman, C. and Hagedoorn, J. (1992), 'Convergence and divergence in the internationalisation of technology', paper for MERIT Conference, University of Limburg, December

Freeman, C., Harlow, C. J. E. and Fuller, J. K. (1965), 'Research and development in electronic capital goods', *National Institute Economic Review*, 34: 40–97

Freeman, C. and Lundvall, B.A. (eds.) (1988), *Small Countries Facing the Technological Revolution*, London: Pinter

Freeman, C. and Oldham, G. (1992), 'Requirements for science and technology policy in the 1990s', in Freeman, C. (ed.), *The Economics of Hope*, London: Pinter

Freeman, C. and Perez, C. (1988), 'Structural crises of adjustment: business cycles and investment behaviour', in Dosi, G. *et al.* (eds.), *Technical Change and Economic Theory*, London: Frances Pinter; New York: Columbia University Press

Freeman, C., Sharp, M. and Walker, W. (eds.) (1991), *Technology and the Future of Europe: Global Competition and the Environment in the 1990s*, London and New York: Pinter

Freeman, C. and Soete, L. (1997), *The Economics of Industrial Innovation and Technological Change*, London: Pinter

Freeman, C. and Soete, L. (eds.) (1990), *New Explorations in the Economics of Technical Change*, London: Pinter

Friedman, D. B. and Samuels, R. J. (1992), 'How to succeed without really flying: the Japanese aircraft industry and Japan's technology policy', Cambridge, Mass.: MIT Japan Program

Friedman, M. (1953), 'The methodology of positive economics', *Essays in Positive Economics*, Chicago: Chicago University Press

Gaffard, J-L. (1991), *Economie industrielle de l'innovation*, Paris: Dalloz

Gann, D. (1992), 'Intelligent Building Technologies: Japan and Singapore', DTI Overseas Science and Technology Mission, Visit Report, London: DTI
(1993), *Innovation and the Built Environment*, Brighton: University of Sussex, SPRU

Gazis, D. L. (1979), 'The influence of technology on science: a comment on some experiences of IBM research', *Research Policy*, 8(4): 244–59

Geroski, P., Machin, S. and van Reenen, J. (1993), 'The profitability of innovating firms', *RAND Journal of Economics*, 24: 198–211

Gerschenkron, A. (1962), *Economic Backwardness in Historical Perspective*, Cambridge, Mass., Belknap Press

Gershuny, J. (1983), *Social Innovation and the Division of Labour*, Oxford: Oxford University Press

Gershuny, J. and Miles, I. D. (1983), *The New Service Economy*, London: Pinter

Gerstenberger W. (1990), 'Reshaping industrial structures', Technology/Economy Programme (TEP), International Conference on 'Technology and Competitiveness', Paris, 25–27 June

Gibbons, M. and Johnston, R. (1974), 'The role of science in technological innovation', *Research Policy*, 3: 220–42

Gibbons, M. and Metcalfe, J. S. (1986), 'Technological variety and the process of competition', in Arcangeli, F., David, P. and Dosi, G. (eds.), *Frontiers of Innovation Diffusion*, Report of the Venice Conference on Innovation Diffusion, DAEST, Venice

Gilfillan, S. C. (1935), *The Sociology of Invention*, Chicago: Follet Publishing Company

Gille, B. (1978), *Histoire des Techniques*, Paris: Gallimar

Gjerding, A. N., Johnson, B., Kallehauge, L., Lundvall, B-Å. and Madsen, P. T. (1992), *The Productivity Mystery: Industrial Development in Denmark in the Eighties*, Copenhagen: DJØF Publishing

Glismann, H.H. and Horn, E.-J. (1988), 'Comparative invention performance of major industrial countries: patterns and explanations', *Management Science*, 34: 1169–87

Goddard, J., Thwaites, A. and Gibbs, D. (1986), 'The regional dimension to technological change in Great Britain', in Amin, A. and Goddard, J. B. (eds.), *Technological Change, Industrial Restructuring and Regional Development*, London: Allen and Unwin

Gold, B. (1981), 'Technological diffusion in industry: research needs and shortcomings', *Journal of Industrial Economics*, 29(3): 247–69

Gomulka, S. (1990), *The Theory of Technological Change and Economic Growth*, London: Routledge

Goodwin, R. M. (1951), 'The nonlinear accelerator and the persistence of business cycles', *Econometrica*, 19: 1–17

Gordon, R.J. (1990), *The Measurement of Durable Goods Prices*, Chicago: University of Chicago Press

Goto, A. (1982), 'Business groups in a market economy', *European Economic Review*, 19: 53–70

Gowing, M. (1964), *Britain and Atomic Energy: Vol. I, Independence and Deterrence, Vol. II, Policy Execution*, London: Macmillan

Granstrand, O. (1982), *Technology, Management and Markets*, London: Pinter

(1986), 'The modelling of buyer/seller diffusion processes. A novel approach to modelling diffusion and simple evolution of market structure', in Arcangeli, F., David, P. and Dosi, G. (eds.), *Frontiers of Innovation Diffusion*, Report of the Venice Conference on Innovation Diffusion, DAEST, Venice

Granstrand, O. and Sjölander, S. (1992), 'Managing innovation in multi-technology corporations', *Research Policy*, 19(1): 35–61

Graves, A. (1987), 'Comparative trends in automobile R&D', DRC Discussion Paper 54, Brighton: SPRU, University of Sussex

(1992), 'International competitiveness and technological development in the world automobile industry', D.Phil. thesis, Brighton: University of Sussex

Greenaway D. and Milner C. (1986), *The Economics of Intra-industry Trade*, Oxford: Basil Blackwell

Greenstein, S. (1990), 'Creating economic advantage by setting compatibility standards: can physical tie-in extend monopoly power?, *Economics of Innovation and New Technology*, 1(1): 63–85

Griliches, Z. (1958), 'Research costs and social returns: hybrid corn and related innovations', *Journal of Political Economy*, 66(5): 419–31

(1979), 'Issues in assessing the contribution of research and development to productivity growth', *Bell Journal of Economics*, 10: 92–116

(1980), 'R&D and the productivity slowdown, *American Economic Review*, 70: 343–8

(1986), 'Productivity, R&D and basic research at the firm level in the 1970s', *American Economic Review*, 76: 141–54

(1989), 'Patents: recent trends and puzzles', *Brookings Papers on Economic Activity: Microeconomics*: 291–330

(1990a), 'Patent statistics as economic indicators: a survey', *Journal of Economic Literature*, 28: 1661–707

(1990b), 'The search for R&D spillovers', *Harvard University Working Papers*, Cambridge, Mass.

(1995), 'R&D and productivity: econometric results and measurement issues', in Stoneman, P. (ed.), *Handbook of the Economics of Innovation and Technological Change*, Oxford: Blackwell

Griliches, Z. (ed.) (1984), *R&D, Patents and Productivity*, Chicago: University of Chicago Press

Griliches, Z. and Lichtenberg, F. (1984), 'R&D and productivity growth at the industry level: is there still a relationship?', in Griliches, Z. (ed.), *R&D, Patents and Productivity*, Chicago: University of Chicago Press

Grossman G.M. and Helpman E., (1991), *Innovation and Growth in the Global Economy*, Cambridge, Mass.: MIT Press

Grübler, A. (1990), *The Rise and Fall of Infrastructures*, Heidelberg: Physica-Verlag

Grupp, H. (1990), 'On the supplementary functions of science and technology indicators', *Scientometrics*, 19: 447–72

(1991), 'Innovation dynamics in OECD countries: towards a correlated network of R&D-intensity, trade, patent and technometric indicators', in OECD (ed.), *Technology and Productivity: the Challenge for Economic Policy*, Paris: OECD

(1994a), 'The measurement of technical performance of innovations by technometrics and its impact on established technology indicators', *Research Policy*, 23(2): 175–93

(1994b), 'Science, high technology and the competitiveness of EC nations', *Cambridge Journal of Economics*, 19: 209–23

(1996), 'Spillover effects and the science base of innovations reconsidered: an empirical approach', *Journal of Evolutionary Economics*, 6(2): 175–97

(ed.) (1992), *Dynamics of Science-Based Innovation*, Berlin: Springer Verlag

Grupp, H. and Hohmeyer, O. (1986), 'A technometric model for the assessment of technological standards and their application to selected technology comparisons', *Technological Forecasting and Social Change* 30: 123–37

Grupp, H., Maital, S., Frenkel, A. and Koschatzky, K. (1992), 'A data envelopment model to compare technological excellence and export sales in Israel and European Community countries', *Research Evaluation*, 2(2): 87–101

Grupp, H. and Soete, L. (1993), 'Analysis of the dynamic relationship between technical and economic performances in ICT sectors 1', Synthesis Report to EC; DGXIII, Karlsruhe: ISI and Maastricht: MERIT

Guerrieri, P. (1991), 'Technology and international trade performance of the most advanced countries', BRIE Working Papers, University of California, Berkeley

(1992), 'Technological and trade competition: the case of US, Japan and Germany', in Harris, M. and Moore, G. E. (eds.), *Linking Trade and Technology Policies*, Washington, DC: National Academy of Engineering

(1993), 'Patterns of technological capability and international trade performance: an empirical analysis', in Kreinin, M. (ed.), *The Political Economy of International Commercial Policy: Issues for the 1990s*, London: Taylor and Francis

Guerrieri P. and Milana C. (1990), *L'Italia e il commercio mondiale*, Bologna: Il Mulino

—— (1991), 'Technological and trade competition in high-tech products', *BRIE Workings Papers* 54, University of California, Berkeley

—— (1995), 'Changes and trends in the world trade in high-technology products', *Cambridge Journal of Economics*, 19(1): 225–42

Hagedoorn, J. (1990), 'Organisational modes of interfirm cooperation and technology transfer', *Technovation*, 10(1): 17–30

Hagedoorn J. and Schakenraad, J. (1990), 'Strategic partnering and technological cooperation', in Freeman, C. and Soete, L. (eds.), *New Explorations in the Economics of Technical Change*, London: Pinter

—— (1992), 'Leading companies and networks of strategic alliances in information technologies', *Research Policy*, 21(2): 163–91

Hahn, F. (1987), 'Information dynamics and equilibrium', *Scottish Journal of Political Economy*, 34(4), 321–34

Håkanson, H. (1989), *Corporate Technological Behaviour: Cooperation and Networks*, London: Routledge

Håkanson, H. and Johansson, J. (1988), 'Formal and informal cooperation strategies in international industrial networks', in Contractor, F. J. and Lorange, P. (eds.), *Cooperative Strategies in International Business*, Lexington. Mass.: Lexington Books

Hamberg, D. (1964), 'Size of firm, oligopoly and research: the evidence', *Canadian Journal of Economic and Political Science*, 30(1): 62–75

—— (1966), *Essays on the Economics of Research and Development*, New York: Random House

Hanusch, H. (ed.) (1988), *Evolutionary Economics: Applications of Schumpeter's Ideas*, Cambridge: Cambridge University Press

Harris, R. I. D. (1988), 'Technological change and regional development in the UK: evidence from the SPRU database', *Regional Studies*, 22: 361–74

Haudeville, B. and Humbert, M. (eds.), *Performances technologiques et performances économiques*, Paris: Presses Univérsitaires de France

Hawkins, R. (1992), 'Standards for technologies of communication: policy implications of the dialogue between technical and non-technical factors', D.Phil. thesis, Brighton: University of Sussex, SPRU

Heertje, A. (1977), *Economic and Technical Change*, London: Weidenfeld and Nicolson

—— (1992), 'Capitalism, socialism and democracy after fifty years', International Economics Association, Tenth World Congress, Moscow

Heertje, A. (ed.) (1988), *Innovation, Technology and Finance*, Oxford: Blackwell

Heertje, A. and Perlman, M. (eds.) (1990), *Evolving Technology and Market Structure*, Ann Arbor: University of Michigan Press

Heiner, R. (1983), 'The origin of predictable behavior', *American Economic Review*, 73(4): 560–95

(1988), 'Imperfect decisions, routinized behavior and inertial technical change', in Dosi, G. *et al.* (eds.), *Technical Change and Economic Theory*, London: Pinter, New York: Columbia University Press

Hessen, B. (1931), 'The social and economic roots of Newton's *Principia*', in Bukharin, N. (ed.), *Science at the Crossroads*, reprinted 1971, London: Frank Cass

Hicks, D. *et al.* (1992a), 'Japanese corporations, scientific research and globalisation', *DRC Discussion Paper 91*, ESRC Research Centre, Brighton: University of Sussex, SPRU

Hicks, D., Isard, P. and Hirooka, M. (1992b), 'Science in Japanese companies', *Japan Journal for Science, Technology and Society*, 1: 108–49

Hill, C. and Utterback, J. (1979), *Technological Innovation for a Dynamic Economy*, Oxford: Pergamon

Hilpert, U. (1991), *State Policies and Techno-industrial Innovation*, London: Routledge

Hirschman, A. O. (1958), *The Strategy of Economic Development*, New Haven, CT: Yale University Press

Hobday, M. (1992), 'Foreign investment, exports and technology development in the four dragons', UN TNC Division, Campinas Conference, Brazil, November

Hodgson, G. M. (1991), 'Evolution and intention in economic theory', in Saviotti, P. and Metcalfe, J. S. (eds.), *Evolutionary Theories of Economic and Technological Change*, Reading: Harwood Academic Publishers

Hodgson, G. M. (1992), 'Optimisation and evolution: Winter's critique of Friedman revisited', Newcastle Polytechnic, Department of Economics

(1993), *Economics and Evolution: Bringing Life Back into Economics*, Cambridge: Polity Press

Hoffmann, K. and Rush H. (1988), *Microelectronics and the Clothing Industry*, New York: Praeger

Holbek, J. (1988), 'The innovation-design dilemma', in Grønlaug, K. and Kaufmann, E. (eds.), *Innovation: A Cross-Disciplinary Perspective*, Oslo: Norwegian University Press

Hollander, S. (1965), *The Sources of Increased Efficiency: A Study of DuPont Rayon Plants*, Cambridge, Mass.: MIT Press

Hollingsworth, R. (1993), 'Variation among nations in the logic of manufacturing sectors and international competitiveness', in Foray, D. and Freeman, C. (eds.), *Technology and the Wealth of Nations*, London: Pinter, chapter 13

Hood, N. and Vahlne, J. E. (eds.) (1988), *Strategies in Global Competition*, London: Croom Helm

Hounshell, D. A. (1992a), 'DuPont and large-scale R&D', in Galison, P. and Herly, B., *Big Science: The Growth of Large-Scale Research*, Stanford: Stanford University Press

(1992b), 'Continuity and change in the management of industrial research: the DuPont Company 1902–1980', in Dosi, G., Giannetti, R. and Toninelli, P. A. (eds.), *Technology and Enterprise in a Historical Perspective*, Oxford: Oxford University Press

Hounshell, D. A. and Smith, J. K. (1988), *Science and Corporate Strategy: DuPont R&D 1902–1980*, Cambridge: Cambridge University Press

Howell, D. R. and Wolff, E. N. (1991), 'Trends in the growth and distribution of skills in the US workplace, 1960–85', *Industrial and Labor Relations Review*, 44: 486–502

Howells, J. (1990), 'The globalisation of research and development, a new era of change', *Science and Public Policy*, 17(4): 273–85

Hu, Y-S. (1975), 'National Attitudes and the Financing of Industry', *Political and Economic Planning*, 41, Broadsheet No. 559, London

(1992), 'Global or transnational corporations are national firms with international operations', *California Management Review*, 34(2): 107–26

Hufbauer, G. C. (1966), *Synthetic Materials and the Theory of International Trade*, London: Duckworth

Hughes, K. (1986), *Technology and Exports*, Cambridge: Cambridge University Press

Hughes, T. P. (1982), *Networks of Power: Electrification in Western Society 1800–1930*, Baltimore, MD: Johns Hopkins University Press

(1989), *American Genesis*, New York: Viking

(1992), 'The dynamics of technological change: salients, critical problems and industrial revolutions', in Dosi, G., Giannetti, R. and Toninelli, P. A. (eds.), *Technology and Enterprise in a Historical Perspective*, Oxford: Oxford University Press

Imai, K. (1989), 'Evolution of Japan's corporate and industrial networks', in Carlsson, B. (ed.), *Industrial Dynamics*, Boston: Kluwer Academic Publishers

Imai, K. and Baba, Y. (1989), 'Systemic innovation and cross-border networks: transcending markets and hierarchies to create a new techno-economic system', OECD Conference on Science, Technology and Economic Growth, Paris, June

Imai, K. and Itami, H. (1984), 'Interpenetration in organisation and market: Japan's firm and market in comparison with US', *International Journal of Industrial Organisation*, 2: 285–310

International Trade Administration (1985), 'High technology trade and competitiveness', Staff Report, DIE-01-85

Irvine, J., Martin, B. R., Abraham, J. and Peacock, T. (1987), 'Assessing basic research: reappraisal and update of an evaluation of four radio astronomy laboratories', *Research Policy*, 16(2–4): 213–27

Iwai, K. (1984a), 'Schumpeterian dynamics Part I: an evolutionary model of innovation and imitation', *Journal of Economic Behavior and Organisation*, 5: 159–90

(1984b), 'Schumpeterian dynamics, Part II: technological progress, firm growth and economic selection', *Journal of Economic Behavior and organisation*, 5: 321–51

Jacobsson, S. (1986), *Electronics and Industrial Policy: The Case of Computer-Controlled Machine Tools*, London: Allen and Unwin

Jagger, N. S. B. and Miles, I. D. (1991), 'New telematic services in Europe', in Freeman, C., Sharp, M. and Walker, W. (eds.), *Technology and the Future of Europe*, London: Pinter

Jaikumar R. (1988), 'From filing and fitting to flexible manufacturing', Working Paper 88–045, Harvard Business School

Jang-Sup Shin (1992), 'Catching up and technological progress in late industrialising countries', M. Phil. dissertation, Cambridge

Jewkes, J., Sawers, D. and Stillerman, R. (1958), *The Sources of Invention*, London and New York: Macmillan (revised edn 1969)

Johnson, B. (1992), 'Institutional learning', in Lundvall, B-Å. (ed.) *National Systems of Innovation*, London: Pinter

Johnson, C., Tyson, L. and Zysman, J. (eds.), (1990), *Politics and Productivity: How Japan's Development Strategy Works*, New York: Harper Business

Johnston, J. (1984), *Econometric Methods*, New York, McGraw-Hill

Johnston, R. and Hartley, J. (1986), 'Formulation and development of high technology indicators', Centre for Technological and Social Change, University of Wollongong

Jorgenson, D. (1996), 'Endogenizing investment in tangible assets, education, and new technologies', paper presented at the seminar 'Finance, Research, Education, and Growth', University of Rome 'Tor Vergata', Rome, June

Kaldor, N. (1957), 'A new model of economic growth', *Economic Journal*, 67: 591–624

(1978), 'The effects of devaluation on trade', *Further Essays on Economic Policy*, London: Duckworth

(1981), 'The role of increasing returns, technical progress and cumulative causation in the theory of international trade and economic growth', *Economie Appliquée*, 34: 593–617

Kaldor, N. and Mirrlees, J. A. (1962), 'A new model of economic growth', *Review of Economic Studies*, 29: 174–92

Kamien, M. and Schwartz, N. (1975), 'Market structure and innovation: a survey', *Journal of Economic Literature*, 13(1): 1–37

(1982), *Market Structure and Innovation*, Cambridge University Press

Kaplinsky, R. (1983), 'Firm size and technical change in a dynamic context', *Journal of Industrial Economics*, 32: 39–59

Katz, B. G. and Phillips, A. (1982), 'Government, technological opportunities and the emergence of the computer industry', in Giersch, H. (ed.), *Emerging Technologies*, Thibinger: J. C. B. Mohr

Katz, L.F. and Summers,L.H. (1989), 'Industry rents: Evidence and Implications', *Brookings Papers on Economic Activity: Microeconomics*: 209–75

Kay, N. (1979), *The Innovating Firm*, London: Macmillan

(1982), *The Evolving Firm*, London: Macmillan

Keck, O. (1982), *Policy-Making in a Nuclear Reactor Programme: The Case of the West German Fast-Breeder Reactor*, Lexington: Lexington Books

Kelley, M. B. and Brooks, H. (1991), 'External learning opportunities and the diffusion of process innovations to small firms: the case of programmable automation', in Nakicenovic, N. and Grübler, A. (eds.), *Diffusion of Technologies and Social Behaviour*, Berlin: Springer

Kennedy, C. and Thirlwall, A. P. (1973), 'Technical progress', *Surveys in Applied Economics* vol. I, London: Macmillan

Kitson, M. and Michie, J. (1995a), 'Trade and growth: a historical perspective', in Michie, J. and Grieve Smith, J. (eds.), *Managing the Global Economy*, Oxford: Oxford University Press

(1995b), 'Conflict, cooperation and change: the political economy of trade and trade policy', *Review of International Political Economy*, 2(4): 632–57

Kitti, C. and Schiffel, D. (1978), 'Rates of invention: international patent comparisons', *Research Policy*, 7: 323–40

Klein, B. H. (1977), *Dynamic Economics*, Cambridge, Mass.: Harvard University Press

Kleinknecht, A. (1987), *Innovation Patterns in Crisis and Prosperity: Schumpeter's Long Cycle Reconsidered*, London: Macmillan

(1990), 'Are there Schumpeterian waves of innovation?', *Cambridge Journal of Economics*, 14(1): 81–92

Kleinknecht, A. and Reijnen, J. O. N. (1992), 'The experience with new innovation data in the Netherlands', *STI Review*, 11: 64–76

(1992a), 'Why do firms cooperate on R&D? An empirical study', *Research Policy*, 21(4): 347–60

Kline, S. and Rosenberg, N. (1985), 'An overview of the process of innovation', in Landau, R. and Rosenberg, N. (eds.), *The Positive Sum Strategy: Harnessing Technology for Economic Growth*, Washington, DC: National Academy Press

Kodama, F. (1986), 'Japanese innovation in mechatronics technology', *Science and Public Policy*, 13(1): 44–51

(1990), 'Rivals' participation in collective research: economic and technological rationale', Tokyo: NISTEP Conference, 2–4 February

(1991), *Analysing Japanese High Technologies*, London: Pinter

(1992), 'Japan's unique capacity to innovate: technology fusion and its international implications', in Arrison, T. S., Bergsten, C. F., Graham, E. M. and Harris, M. C. (eds.), *Japan's Growing Technological Capability: Implications for the US Economy*, Washington, DC: National Academy Press

Krauch, H. (1970), *Die organisierte Forschung*, Berlin: Luchterhand

(1990), *Prioritäten für die Forschungspolitik*, Munich: Hanser Verlag

Kreinin, M. E., (1993), 'Super-301 and Japan: a dissenting view', in Kreinin, M. (ed.), *The Political Economy of International Commercial Policy: Issues for the 1990s*, London and New York: Taylor & Francis

Krugman, P. (1981), 'Trade, accumulation and uneven development', *Journal of Development Economics*, 8: 149–161

Krugman, P. (1987), 'The narrow moving band, the Dutch disease, and the competitive consequences of Mrs. Thatcher', *Journal of Development Economics*, 27: 41–55

(1990), *Rethinking International Trade*, Cambridge, Mass.: MIT Press

(1991a), *Geography and Trade*, Cambridge, Mass.: MIT Press

(1991b), 'Myths and realities of US competitiveness', *Science*, 254: 811–15

(1992), 'Technology and international competition: a historical perspective', in Harris, M. and Moore, G. E. (eds.), *Linking Trade and Technology Policies*, Washington, DC: National Academy of Engineering

(1995), 'Technological change in international trade', in Stoneman, P. (ed.),

Handbook of the Economics of Innovation and Technological Change, Oxford: Blackwell

Krupp, H. (1992), *Energy Politics and Schumpeter Dynamics*, Tokyo: Springer

Kuhn, T. S. (1962), *The Structure of Scientific Revolutions*, Chicago: Chicago University Press

Kuznets, S. (1973), *Population, Capital and Growth: Selected Essays*, New York: W.W. Norton and Company

Lall, S. (1987), *Learning to Industrialise: The Acquisition of Technological Capability by India*, London: Macmillan

Landes, M. (1970), *The Unbound Prometheus: Technological and Industrial Development in Western Europe from 1750 to the Present*, Cambridge: Cambridge University Press

Langrish, J., Gibbons, M., Evans, P. and Jevons, F. (1972), *Wealth from Knowledge*, London: Macmillan

Lastres, H. (1992), 'Advanced materials and the Japanese national system of innovation', D.Phil. thesis, Brighton: University of Sussex, SPRU

Lawrence, P. (1980), *Managers and Management in W. Germany*, London: Croom Helm

Lawrence, R. Z. (1991), 'Efficient or exclusionist? The import behavior of Japanese corporate groups', *Brookings Papers on Economic Activity*, 1: 311–30

Lazonick, W. (1990), *Competitive Advantage on the Shop Floor*, Cambridge, Mass.: Harvard

(1992a), 'Business organisation and competitive advantage: capitalist transformations in the twentieth century', in Dosi, G., Giannetti, R. and Toninelli, P. A. (eds.), *Technology and Enterprise in a Historical Perspective*, Oxford: Oxford University Press

(1992b), 'Controlling the market for corporate control: the historical significance of managerial capitalism', in Scherer F. M. and Perlman, M. (eds.), *Entrepreneurship, Technological Innovation and Economic Growth: Studies in the Schumpeterian Tradition*, Ann Arbor: University of Michigan Press

Legler, H. (1987), 'West German competitiveness of technology-intensive products', in Grupp, H., (ed.), *Problems of Measuring Technological Change*, Cologne: Verlag TUEV

(1990), 'The German competitive position in trade of technology-intensive products', in G. Heiduk and K. Yamamura (eds.), *Technological Competition and Interdependence*, Seattle and London: Washington University Press

Legler, H., Grupp, H., Gehrke, B. and Schasse, U. (1992), *Innovationspotential und Hochtechnologie*, Heidelberg and New York: Physica-Springer

Levin, R. C. (1986), 'A new look at the patent system', *American Economic Review*, 76: 199–201

(1988), 'Appropriability, R&D spending and technological performance', *American Economic Review Papers & Proceedings*, 78: 424–8

Levin, R., Cohen, W. M. and Mowery, D. C. (1985), 'R&D appropriability, opportunity and market structure: new evidence on some Schumpeterian hypotheses', *American Economic Review*, 75(2): 20–4

Levin, R. C., Klevorick, A. K., Nelson, R. R. and Winter, S. G. (1987),

'Appropriating the returns from industrial research and development', *Brookings Papers on Economic Activity*, 3: 783–820

Levin, R. *et al.* (1984), *Survey Research on R&D Appropriability and Technological Opportunity*, New Haven: Yale University Press

Levine, R. and Renelt, D. (1992), 'A sensitivity analysis of cross-country growth regressions', *American Economic Review*, 82: 942–63

Lichtenberg, F.R. (1992), 'R&D investment and international productivity differences', Cambridge, Mass., NBER Working Paper No. 4161, September

Lichtenberg, F.R. and Siegel, D. (1991), 'The impact of R&D investment on productivity: new evidence using linked R&D–LRD data,' *Economic Inquiry*, 29: 203–28

Limpens, I., Verspagen, B. and Beelen, E. (1992), *Technology Policy in Eight European Countries: A Comparison*, University of Limberg, MERIT

Linder, S. B. (1961), *An Essay on Trade and Transformation*, Uppsala: Almquist & Wicksell

List, F. (1841), *Das Nationale System der Politischen Ökonomie*, Stuttgart and Tubingen: J. G. Cotta

Lorenz, D. (1967), *Dynamische Theorie der internationalen Arbeitsteilung*, Berlin: Duncker & Humblot

Lovio, R. (1993), 'Evolution of firm communities in new industries: the case of the Finnish electronics industry', Helsinki, Ph.D. Dissertation, School of Economics and Business Administration

Lucas, R. E. B. (1986), 'Adaptive behaviour and economic theory', *Journal of Business*, 59: 401–76

Lundgren, A. (1991), 'Technological innovation and industrial evolution: the emergence of industrial networks', D.Phil. dissertation, Stockholm School of Economics

Lundvall, B-Å. (1985), 'Product innovation and user-producer interaction', *Industrial Development Research Series*, 31, Aalborg: Aalborg University Press

(1988), 'Innovation as an interactive process: from user-producer interaction to the national system of innovation', in Dosi, G. *et al.* (eds.), *Technical Change and Economic Theory*, London: Pinter

(1993), 'User–producer relationships, national systems of innovation and internationalisation', in Foray, D. and Freeman, C. (eds.), *Technology and the Wealth of Nations*, London: Pinter

(1996), 'The learning economy. Challenges to economic theory and policy', paper presented at the Euro-Conference 'National Systems of Innovation or the Globalisation of Technology?', National Research Council, Rome

Lundvall, B-Å. (ed.) (1992), *National Systems of Innovation: Towards a Theory of Innovation and Interactive Learning*, London: Pinter

Machlup, F. (1962), *The Production and Distribution of Knowledge in the United States*, Princeton: Princeton University Press

Mackenzie, D. (1990a), *Inventing Accuracy: A Historical Sociology of Nuclear Missile Guidance*, Cambridge, Mass.: MIT Press

(1990b), 'Economic and sociological exploration of technical change', paper presented at the Manchester Conference on Firm Strategy and Technical Change

Mackerron, G. S. (1991), 'De-commissioning costs and British nuclear policy', *Energy Policy*, 12: 13–28

MacQueen, D. H. and Wallmark, J. T. (1983), *100 Viktige Innovationer i Sverige 1945–1980*, Stockholm: STU

Maddison, A. (1982), *Phases of Capitalist Development*, New York: Oxford University Press

(1991), *Dynamic Forces in Capitalist Development*, New York: Oxford University Press

Magee, S. P. (1975), 'Prices, incomes, and foreign trade', in Kenen, P. S., *International Trade and Finance. Frontiers for Research*, Cambridge: Cambridge University Press

Mahajan, V. and Peterson, R. A. (1979), 'Integrating firm and space in technological diffusion models', *Technological Forecasting and Social Change*, 14: 127–46

Malerba, F. (1985), *The Semiconductor Business: The Economics of Rapid Growth and Decline*, Madison: University of Wisconsin Press

Malerba, F., Torrisi, S. and von Tunzelmann, N. (1991), 'Electronic computers', in Freeman, C. *et al.* (eds.), *Technology and the Future of Europe*, London: Pinter

Mankiw, N. G. (1995), 'The growth of nations', *Brookings Papers on Economic Activity*, 1: 275–326

Mankiw, N. G., Romer, D., and Weil, D. (1992), 'A contribution to the empirics of economic growth', *Quarterly Journal of Economics*, 107: 407–38

Mansell, R. (1988), 'Telecommunication network-based services: regulation and market structure in transition', *Telecommunication Policy*, 12(3): 243–55

(1989), *Technology Network-Based Services: Policy Implications*, Paris: OECD

(1990a), 'Rethinking the telecommunication infrastructure: the new black box', *Research Policy*, 19(6): 507–15

(1990b), 'Multinational relationships: shaping telecommunication markets', in Wilkinson, C. (ed.), *International Aspects of Industrial Policies*, Luxembourg: Institut Universitaire International: 65–80

Mansell, R. and Morgan, K. (1991), 'Evolving telecommunication infrastructures: organising the new European community market-place', in Freeman, C. *et al.* (eds.), *Technology and the Future of Europe*, London: Pinter

Mansfield, E. (1961), 'Technical change and the rate of imitation', *Econometrica* 29(4): 741–66

(1968), *The Economics of Technological Change*, New York: Norton

(1977), *The Production and Application of New Industrial Technology*, New York: Norton

(1980), 'Basic research and productivity increase in manufacturing', *American Economic Review*, 70: 863–73

(1985), 'How rapidly does new industrial technology leak out?', *Journal of Industrial Economics*, 34(2): 217–23

(1988), 'Industrial innovation in Japan and the United States', *Science*, 241: 1760–4

(1989), 'The diffusion of industrial robots in Japan and the United States', *Research Policy*, 18: 183–92

(1991), 'Academic research and industrial innovation', *Research Policy*, 20(1): 1–13

Mansfield, E. *et al.* (1971), *Research and Innovation in the Modern Corporation*, New York: Norton, and London: Macmillan

Mansfield, E., Schwartz, M. and Wagner, S. (1981), 'Imitation costs and patents: an empirical study', *Economic Journal*, 91(4): 907–18

March, J. G. and Simon, H. A. (1958), *Organisations*, New York: Wiley

Marsden, D. (1993), 'Skill flexibility, labour market structure, training systems and competitiveness', in Foray, D. and Freeman, C. (eds.), *Technology and the Wealth of Nations*, London: Pinter

Marshall, A. (1890), *Principles of Economics*, London: Macmillan

Martin, B. R. and Irvine, J. (1981), 'Internal criteria for scientific choice: an evolution of the research performance of electric high-energy physics accelerators', *Minerva*, 19: 408–32

—— (1983), 'Assessing basic research: some partial indicators of scientific progress in radio astronomy', *Research Policy*, 12: 61–90

—— (1985), 'Evaluating the evaluators: a reply to our critics', *Social Studies of Science*, 15: 558–75

Mason, G., Prais, S. and van Ark, B. (1992), 'Vocational education and productivity in the Netherlands and Britain', *National Institute Economic Review*, May: 45–63

Matthews, J. (1989), *Tools of Change: New Technology and the Democratisation of Work*, Sydney: Pluto Press

Meliciani, V. and Simonetti, R. (1996), 'Technical change, patterns of specialisation and economic growth', paper presented at the Conference The Multiple Linkages between Technological Change and the Economy, University of Rome, 'Tor Vergata', March

Mensch, G. (1975), *Das technologische Patt*, Frankfurt: Umschau

—— (1979), *Stalemate in Technology*, Cambridge, Mass.: Ballinger

Metcalfe, J. S. (1981), 'Impulse and diffusion in the study of technical change', *Futures*, 13(5): pp 347–59

—— (1988), 'The diffusion of innovations: an interpretative survey' in Dosi, G. *et al.* (eds.), *Technical Change and Economic Theory*, London: Pinter

—— (1995a), 'Technology systems and technology policy in an evolutionary framework', *Cambridge Journal of Economics*, 19(1): 25–46

—— (1995b), 'The economic foundations of technology policy: equilibrium and evolutionary perspectives', in Stoneman, P. (ed.), *Handbook of the Economics of Innovation and Technical Change*, Oxford: Blackwell

Metcalfe, J. S. and Gibbons, M. (1983), 'On the economics of structural change and the evolution of technology', Manchester University paper presented at the 7th World Congress of the International Economics Association, Madrid, September (1987), in L. Pasinetti (ed.), *Growth and Structural Change*, London: Macmillan

Meyer-Krahmer, F. (1990), *Science and Technology in the Federal Republic of Germany*, London: Longman

—— (1992), 'The German R&D system in transition: empirical results and prospects of future development', *Research Policy*, 21(5): 423–37

Meyer-Krahmer, F. and Soligny, P. (1989), 'Evaluation of government innovation programmes in selected European countries', *Research Policy*, 18: 315–32

Midgley, D. F., Morrison, P. D. and Roberts, J. H. (1992), 'The effect of network structure in industrial diffusion processes', *Research Policy*, 21(6): 533–52

Milana, C., (1988), 'Constant-market-shares analysis and index number theory', *European Journal of Political Economy*, 4: 453–78

Miles, I. (1989), *Home Informatics: Information Technology and the Transformation of Everyday Life*, London: Pinter
    (1990), 'Teleshopping: just around the corner?', *Journal of the RSA*, 138: 180–9

Miles, I. and Thomas, G. (1990), 'The development of new telematics services', *STI Review*, 7: 35–63

Miles, I., Rush, H., Turner K. and Bessant, J. (1988), *Information Horizons: The Long-Term Social Implications of new IT*, London: Edward Elgar

Miles, I., Schneider, V. and Thomas, G. (1991), 'The dynamics of videotex development in Britain, France and Germany: a cross-national comparison', *European Journal of Communication*, 6: 187–212

Miller, R., Hobday, M., Leroux-Demers, T. and Olleros, X. (1993), 'Innovation in complex systems industries: the case of flight simulation', *Industrial and Corporate Change*, 3(4)

Mitchell, G. and Hamilton, W. (1988), 'Managing R&D as a strategic option', *Research-Technology Management*, 31: 15–22

Mjøset, L. (1992), *The Irish Economy in a Comparative Institutional Perspective*, Dublin: National Economic and Social Council

Mohnen, P. (1990), 'New technologies and interindustry spillovers', *STI Review*, no. 7: 131–47
    (1992), *The Relationship between R&D and Productivity Growth in Canada and Other Major Industrialised Countries*, Ottawa: Canada Communications Group

Molina, A. H. (1989), 'Transputers and parallel computers: building technological competition through socio-technical constituencies', PICT paper 7, Edinburgh: Research Centre for Social Services
    (1990), 'Transputers and transputer-based parallel computers', *Research Policy* 19(4): 309–35

Morgan, K. and Sayer, A. (1988), *Microcircuits of Capital*, Cambridge: Polity Press

Morris, P. J. T. (1982), 'The development of acetyline chemistry and synthetic rubber by IG Farben AG 1926–1945', D.Phil. thesis, University of Oxford

Morton, J. A. (1971), *Organizing for Innovation*, New York: McGraw-Hill

Mowery, D. C. (1980), 'The emergence and growth of industrial research in American manufacturing 1899–1946', Ph.D. dissertation, Stanford University
    (1983), 'The relationship between intrafirm and contractual forms of industrial research in American manufacturing 1900–1940', *Explorations in Economic History*, 20(4): 351–74
    (1989), 'Collaborative ventures between US and foreign manufacturing firms', *Research Policy*, 18(1): 19–33
    (1992a), 'Finance and corporate evolution in firm industrial economics 1900–1950', *Industrial and Corporate Change*, 1(1): 1–37
    (1992b), 'The US national innovation system: origins and prospects for change', *Research Policy*, 21(2): 125–45

(1995), 'The practice of technology policy', in Stoneman, P. (ed.), *Handbook of the Economics of Innovation and Technological Change*, Oxford: Blackwell

Mowery, D. C. (ed.) (1988), *International Collaborative Ventures*, Cambridge: Ballinger

Mowery, D. C. and Rosenberg, N. (1979), 'The influence of market demand upon innovation: a critical review of some recent empirical studies', *Research Policy*, 8: 102–53

(1989), *Technology and the Pursuit of Economic Growth*, Cambridge: Cambridge University Press

Münt, G. 1996. *Dynamik von Innovation und Aussenhandel*, Heidelberg, New York: Physica-Springer

Mulder, K. F. and Vergragt, P. J. (1990), 'Synthetic fibre technology and company strategy', *R&D Management*, 20(3): 247–56

Myers, S. (1984), 'Finance theory and finance strategy', *Interfaces*, 14: 126–137

Myers, S. and Marquis, D. G. (1969), *Successful Industrial Innovation*, Washington, DC: National Science Foundation

Nabseth, L. and Ray, G. F. (1974), *The Diffusion of New Industrial Processes*, Cambridge University Press

Nadiri, M.I. (1993), 'Innovation and technological spillovers', Cambridge, Mass., NBER Working Paper No. 4423, August

Nakicenovic, N. and Grübler, A. (eds.) (1991), *Diffusion of Technologies and Social Behaviour*, Berlin: Springer Verlag

Narin, F. and Noma, E. (1985), 'Is technology becoming science?', *Scientometrics*, 7(3): 369–81

Narin, F., Noma, E. and Perry, R. (1987), 'Patents as indicators of corporate technological strength', *Research Policy*, 16(2–4): 143–57

Narin, F. and Olivastro, D. (1992), 'Status report: linkage between technology and science', *Research Policy*, 21: 237–51

National Science Foundation (1969), Report on Project TRACES by Illinois Institute of Technology Research Institute, Washington, DC

Nelson, R. R. (1959a), 'The simple economics of basic scientific research', *Journal of Political Economy*, 67(3): 297–306

(1959b), 'The economics of invention: a survey of the literature', *Journal of Business*, 32(2): 101–27

(1962), 'The link between science and invention: the case of the transistor', in Nelson, R. (ed.), *The Rate and Direction of Inventive Activity*, Princeton: Princeton University Press

(1981), 'Research on productivity growth and differences', *Journal of Economic Literature*, 19: 1029–64

(1984), *High Technology Policies: A Five-Nation Comparison*, Washington, DC: American Enterprise Institute

(1985), 'Institutions supporting technical advance in industry', *American Economic Review*, 85: 186–9

(1987), *Understanding Technical Change as an Evolutionary Process*, Amsterdam: North-Holland

(1988), 'Institutions supporting technical change in the United States', in Dosi, G. et al. (eds.), *Technical Change and Economic Theory*, London: Pinter

(1990a), 'Capitalism as an engine of progress', *Research Policy*, 19: 193–214

(1990b), 'US technological leadership: where did it come from, and where did it go?', *Research Policy*, 19: 117–32

(1990c), 'What has happened to US technological leadership?', in Heiduk, G. and Yamamura, K. (eds.), *Technological Competition and Interdependence*, Seattle and London: Washington University Press

(1991), 'Why do firms differ, and how does it matter?', *Strategic Management Journal*, 12(1)

(1992a), 'What is "commercial" and what is "public" about technology, and what should be', in Rosenberg, N., Landau, R. and Mowery, D. (eds.), *Technology and the Wealth of Nations*, Stanford: Stanford University Press

(1992b), 'National innovation systems: a retrospective on a study', *Industrial and Corporate Change*, 1(2): 347–74

(1992c), 'The roles of firms in technical advance: a perspective from evolutionary theory', in Dosi, G., Giannetti, R. and Toninelli, P. A. (eds.), *Technology and Enterprise in a Historical Perspective*, Oxford: Oxford University Press

Nelson, R. R. (ed.) (1962), *The Rate and Direction of Inventive Activity*, National Bureau of Economic Research, Princeton: Princeton University Press

(ed.) (1993), *National Innovation Systems: A Comparative Study*, Oxford, Oxford University Press

Nelson, R. R., Peck, M. J. and Kalachek, E. D. (1967), *Technology, Economic Growth and Public Policy*, London: Allen and Unwin

Nelson, R. R. and Phelps, E. S. (1966), 'Investment in humans, technological diffusion and economic growth,' *American Economic Review*, 61: 69–75

Nelson, R. R. and Winter, S. G. (1974), 'Neoclassical versus evolutionary theories of economic growth', *Economic Journal*, 84: 886–905

(1977), 'In search of a useful theory of innovation', *Research Policy*, 6(1): 36–76

(1982), *An Evolutionary Theory of Economic Change*, Cambridge, Mass.: Harvard University Press

Nelson, R. and Wright, G. (1992), 'The rise and fall of American technological leadership: the postwar era in historical perspective', *Journal of Economic Literature*, 30: 1931–64

Newton, K., de Broucker, P., McDougal, G., McMullen, K., Schweitzer, T. and Siedule, T. (1992), *Education and Training in Canada*, Ottawa, Canada Communication Group

Niosi, J. (ed.) (1991), *Technology and National Competitiveness*, Montreal, McGill-Queens University Press

Oakey, R. (1984), *High Technology Small Firms*, London: Pinter

OECD (1963, 1970, 1976, 1981, 1993), *The Measurement of Scientific and Technical Activities*, Paris: Directorate for Scientific Affairs

(1985), *Trade in High Technology Product*, Directorate for Science, Technology and Industry, Paris

(1986), *Technological Agreements between Firms*, Paris: OECD

(1988), 'A survey of technology licensing', *STI Review*, 4: 7–51

(1991a), *TEP: Technology in a Changing World*, Paris: OECD

(1991b), *Technology and Productivity: The Challenges for Economic Policy*, Paris: OECD

(1991c), *Basic Science and Technology Statistics*, Paris: OECD

(1992a), *Technology and the Economy: The Key Relationships*, Paris: OECD

(1992b), 'Special Issue on Innovation Measurement', *STI Review*, 11, Paris: OECD

OECD Economic Policy Committee (1993), 'Medium-term productivity performance in the business sector: trends, underlying determinants and policies', Paris: OECD, September

Ohmae, K. (1990), *The Borderless World*, New York: Harper

Okimoto, D.I., (1989), *Between MITI and the Market: Japanese Industrial Policy for High Technology*, Stanford, CA: Stanford University Press

Olson, M. (1982), *The Rise and Decline of Nations*, New Haven: Yale University Press

Orsenigo, L. (1989), *The Emergence of Biotechnology: Institutions and Markets in Industrial Innovation*, London: Pinter

(1993), 'The dynamics of competition in a science-based technology: the case of biotechnology', in Foray, D. and Freeman, C. (eds.), *Technology and the Wealth of Nations*, London: Pinter

Ostry, S. (1990), *Governments and Corporations in a Shrinking World*, Council on Foreign Relation Book, New York–London

Pack, H. (1988), 'Industrialisation and trade', in Chenery, H. and Srinivasan, T.N. (eds.), *Handbook of Development Economics*, vol. I, Amsterdam: Elsevier Science Publishers B.V

Pasinetti, L. L. (1981), *Structural Change and Economic Growth*, Cambridge: Cambridge University Press

Patel, P. (1995), 'Localised production of technology for global markets', *Cambridge Journal of Economics*, 19(1): 141–53

Patel, P. and Pavitt, K. (1991a), 'Large firms in the production of the world's technology: an important case of 'non-globalisation'', *Journal of International Business Studies*, 22(1): 1–21

(1991b), 'Europe's technological performance', in Freeman, C., Sharp, M. and Walker, W. (eds.), *Technology and the Future of Europe*, London and New York: Pinter

(1992a), 'The innovative performance of the world's largest firms: some new evidence', *Economics of Innovation and New Technology*, 2: 91–102

(1992b), 'The continuing widespread (and neglected) importance of improvements in mechanical technologies', paper presented at Stanford Conference on 'The Role of Technology in Economics' in honour of Nathan Rosenberg, Brighton: University of Sussex, SPRU, mimeo, November

(1994), 'National innovation systems: why they are important and how they might be measured and compared', *Economics of Innovation and New Technology*, 3: 77–95

(1995), 'Patterns of technological activity: their measurement and interpretation', in Stoneman, P. (ed.), *Handbook of the Economics of Innovation and Technological Change*, Oxford: Blackwell

Patel, P. and Soete, L. (1988), 'Measuring the economic effects of technology', *STI Review*, 4: 121–66

Paulinyi, A. (1982), 'Der technologietransfer für die Metallbearbeitung und die preussische Gewerbeförderung 1820–1850', in Blaich, F. (ed.), *Die Rolle des Staates für die wirtschaftliche Entwicklung*, Berlin: Blaich

(1989), *Industrielle Revolution: vom Ursprung der modernen Technik*, Hamburg Deutsches Museum, Rowohlt

Pavitt, K. (1971), *The Conditions for Success in Technological Innovation*, Paris: OECD

(1982), 'R&D, patenting and innovation activities: a statistical exploration', *Research Policy*, 11(1): 35–51

(1984), 'Sectoral patterns of technical change', *Research Policy*, 13: 343–73

(1985), 'Patent statistics as indicators of innovative activities: possibilities and problems, *Scientometrics*, 7(1–2): 77–99

(1986a), '"Chips" and "Trajectories": how does the semiconductor influence the sources and directions of technical change?', in MacLeod, R. (ed.), *Technology and the Human Prospect*, London: Pinter

(1986b), 'Technology, innovation and strategic management', in McGee, J. and Thomas H. (eds.), *Strategic Management Research: A European Perspective*, New York: Wiley

(1988a), 'International patterns of technological accumulation', in Hood, N. and Vahlne, J. E. (eds.), *Strategies in Global Competition*, London: Croom Helm

(1988b), 'Uses and abuses of patent statistics', in Van Raan, A. (eds.), *Handbook of Quantitative Studies of Science and Technology*, Amsterdam: Elsevier

(1990), 'What we know about the strategic management of technology', *California Management Review*, 32: 3–26

(1993), 'What do firms learn from basic research?' in Foray, D. and Freeman, C. *Technology and the Wealth of Nations*, London: Pinter

Pavitt, K. and Patel, P. (eds.) (1980), *Technical Innovation and British Economic Performance*, London: Macmillan

Pavitt, K. and Patel, P. (1988), 'The International distribution and determinants of technological activities', *Oxford Review of Economic Policy*, 4: 35–55

(1990), 'Sources and directions of technological accumulation in France', *Technology Analysis and Strategic Management*, 2: 3–26

Pavitt, K., Robson, M. and Townsend, J. (1987), 'The size distribution of innovative firms in the UK: 1945–1983', *Journal of Industrial Economics*, 35(3): 297–319

(1989), 'Accumulation, diversification and organisation of technological activities in UK companies, 1945–83', in Dodgson, M. (eds.), *Technology Strategy and the Firm: Management and Public Policy*, Harlow: Longman

Pavitt, K., and Sharp, M. (1993), 'Technology policy in the 1990s: old trends and new realities', *Journal of Common Market Studies*, June

Pavitt, K. and Soete, L. L. G. (1980), 'Innovative activities and export shares: some comparisons between industries and countries', in Pavitt, K. (ed.), *Technical Innovation and British Economic Performance*, London: Macmillan

Pavitt, K. and Walker, W. (1976), 'Government policies towards industrial innovation', *Research Policy*, 5(1): 1–96

Pearce, R. (1990), *The Internationalisation of R&D by Multinational Enterprises*, London: Macmillan

References 257

Pearce, R. D. and Singh, S. (1992), *Globalising Research and Development*, London: Macmillan

Penrose, E. (1952), 'Biological analogies in the theory of the firm', *American Economic Review*, 41: 804–19

——— (1959), *The Theory of the Growth of the Firm*, Oxford: Basil Blackwell

Perez, C. (1983), 'Structural change and the assimilation of new technologies in the economic and social system', *Futures*, 15(5): 357–75

——— (1985), 'Microelectronics, long waves and the world structural change: new perspectives for developing countries', *World Development*, 13 (3): 441–63

——— (1989), 'Technical change, competitive restructuring and institutional reform in developing countries', *World Bank Strategic Planning and Review, Discussion Paper 4*, Washington, DC: World Bank

Perrin, J. (1988), *Comment naissent les techniques?* Paris: Publisud

Perraux, F. (1956), 'Note sur la notion de pâte de croissance', *Economie Appliquée*, 7: 307–320

Petit, P. (1991), 'New technology and measurement of services: the case of financial activities', in OECD (1991b), *Technology and Productivity: The Challenges for Economic Policy*, Paris: OECD

——— (1993), 'Technology and employment: main issues in a context of high unemployment', September, mimeo, Paris, CEPREMAP

Petroski, H. (1989), 'H. D. Thoreau, engineer', *American Heritage of Invention and Technology*, New Haven: Yale University Press

Phillips, A. (1971), *Technology and Market Structure*, Lexington, Mass.: Heath

Pianta, M. (1988), *New Technologies across the Atlantic: US Leadership or European Autonomy?*, Hemel Hempstead: Harvester Wheatsheaf

Pianta, M. and Meliciani, V. (1996), 'Technological specialization and economic performance in OECD countries', *Technology Analysis and Strategic Management*, 8: 157–174

Piore, M. J. (1993), 'The revival of prosperity in industrial economies: technological trajectories, organisational structure, competivity', in Foray, D. and Freeman, C. (eds.), *Technology and the Wealth of Nations*, London: Pinter

Piore, M.J., and Sabel, C.F. (1984), *The Second Industrial Divide*, New York: Basic Books

Poon, A. (1993), *Tourism, Technology and Competitive Strategies*, Wallingford: CAB International

Porter, M. (1990), *The Competitive Advantage of Nations*, New York: Free Press and Macmillan

Posner, M. (1961), 'International trade and technical change', *Oxford Economic Papers*, 13(3): 323–43

Posthuma, A. (1986), 'The internationalisation of clerical work: a study of offshore office services in the Caribbean', Occasional Paper 24, Brighton: University of Sussex, SPRU

Prais, S. J. (1981), 'Vocational qualifications of the labour force in Britain and Germany', *National Institute Economic Review*, 98: 47–59

——— (1987), 'Education for productivity: comparisons of Japanese and English schooling for vocational preparation', *National Institute Economic Review*, 119: 40–56

(1993), 'Economic performance and education: the nature of Britain's deficiencies', Discussion Paper no. 52, National Institute for Economic and Social Research, London

Prais, S. J. and Wagner, K. (1983), 'Some practical aspects of human capital investment: training standards in five occupations in Britain and Germany', *National Institute Economic Review*, 105: 46–65

(1988), 'Productivity and management: the training of foremen in Britain and Germany', *National Institute Economic Review*, 123: 34–47

Price, D. de S. (1984), 'The science/technology relationship, the craft of experimental science and policy for the improvement of high technology innovation', *Research Policy*, 13(1): 3–20

Quinn, J. B. (1986), 'The impacts of technology in the services sector', in Guile, B. and Brooks, H. (eds.), *Technology and Global Industry*, Washington, DC: National Academy Press

Quintas, P. (ed.) (1993), *Social Dimensions of Systems Engineering*, Chichester: Ellis Harwood

Quintella, R. (1993), 'The Relationship between business and technology strategies in the chemical industry', D.Phil. thesis, University of Brighton

Ray, G. F. (1984), *The Diffusion of Mature Technologies*, Cambridge, Cambridge University Press

Reati, A. (1992), 'Are we at the eve of a new long-term expansion induced by technological change?', *International Review of Applied Economics*, 6(3): 249–85

Reekie, W. D. (1973), 'Patent data as a guide to industrial activity', *Research Policy*, 2(3, October): 246–66

Reich, L. S. (1985), *The Making of American Industrial Research: Science and Business at GE and Bell 1876–1926*, Cambridge University Press

Reich, R. B. (1991), *The Work of Nations*, New York: Vintage Books

Ricardo, D. (1911), *The Principles of Political Economy and Taxation*, London: Everyman's Library, J. M. Dent and Sons Ltd

Roberts, E. B. (1991), *Entrepreneurs in High Technology: Lessons from MIT and Beyond*, New York: Oxford University Press

Rogers, E. M. (1961), *Diffusion of Innovations*, New York: Free Press of Glencoe

Romeo, A. A (1975), 'Interindustry and interfirm differences in the rate of diffusion of an innovation', *Review Economic Statistics*, 57(3, August): 311–19

Romer, P. M. (1986), 'Increasing returns and long-run growth', *Journal of Political Economy*, 94(5): 1002–37

(1990), 'Endogenous technological change', *Journal of Political Economy*, 98(2): S71–S102

Roobeek, A. J. M. (1987), 'The crisis in Fordism and the rise of a new technological paradigm', *Futures*, 19: 129–154

Rosario, M. and Schmidt, S. K. (1991), 'Standardisation in the EC: the example of ICT', in Freeman, C. *et al.* (eds.), *Technology and the Future of Europe*, London: Pinter

Rosenberg, N. (1963), 'Technological change in the machine tool industry', *Journal of Economic History*, 23: 414–43

(1972), *Technology and American Economic Growth*, New York: Harper Torch Books

(1976), *Perspectives on Technology*, Cambridge: Cambridge University Press

(1982), *Inside the black box: Technology and economics*, Cambridge: Cambridge University Press

(1990), 'Why do firms do basic research with their own money?', *Research Policy*, 19(2): 165–75

(1992), 'Scientific instrumentation and university research', *Research Policy*, 21(4): 381–90

(1994), *Exploring the Black Box: Technology, Economics and History*, Cambridge: Cambridge University Press

Rosenberg, N. and Birdzell, L.E. (1986), *How the West Grew Rich. The Economic Transformation of the Industrial World*, New York: Basic Books

Rosenberg, N., Landau, R. and Mowery, D. (eds.) (1992), *Technology and the Wealth of Nations*, Stanford: Stanford University Press

Rothwell, R. (1977), 'Innovations in textile machinery', *R&D Management*, 6(3): 131–8

(1991), 'External networking and innovation in small and medium-sized manufacturing firms in Europe', *Technovation*, 11(2): 93–112

(1992), 'Successful industrial innovation: critical factors for the 1990s', SPRU 25th Anniversary, Brighton: University of Sussex. Reprinted in *R&D Management*, 22(3): 221–39

Rothwell, R. and Gardiner, P. (1988), 'Re-innovation and robust design: producer and user benefits', *Journal of Marketing Management*, 3(3): 372–87

Rothwell, R. and Zegveld W. (1982), *Industrial Innovation and Public Policy*, London, Pinter

Rothwell, R. *et al.* (1974), 'SAPPHO Updated', *Research Policy*, 3(5): 258–91

Russo, M. (1985), 'Technical change and the industrial district: the role of inter-firm relations in the growth and transformation of ceramic tile production in Italy', *Research Policy*, 14(6): 329–44

Ruttan, V. (1959), 'Usher and Schumpeter on innovation, invention and technological change', *Quarterly Journal of Economics*, 73(4): 596–606

(1982), *Agricultural Research Policy*, Minneapolis: University of Minnesota Press

Sabel, C. F. (1993), 'Studied trust: building new forms of cooperation in a volatile economy', in Foray, D. and Freeman, C. (eds.), *Technology and the Wealth of Nations*, London: Pinter

Sahal, D. (1977), 'The multi-dimensional diffusion of technology', *Technological Forecasting and Social Change*, 10: 277–98

(1981), *Patterns of Technological Innovation*, New York: Addison-Wesley

(1985), 'Technological guide-posts and innovation avenues', *Research Policy*, 14(2): 61–82

Sako, M. (1992), *Contracts, Prices and Trust: How the Japanese and British Manage Their Subcontracting Relationships*, Oxford: Oxford University Press

Salomon, J.-J. (1985), *Le Gauloise, le Cowboy et le Samurai*, Centre de Perspective et d'Évolution, Paris: CNAM

Salter, W. (1960), *Productivity, Growth and Technical Change*, Cambridge: Cambridge University Press

Samuels, R. J. (1987), *Energy Markets in Comparative and Historical Perspective*, Ithaca: Cornell University Press

Santa Fé Institute (1990, 1991, 1992, 1993), Working Papers, Economics Research Programme, New Mexico

Saviotti, P. P. (1991), 'The role of variety in economic and technological development', in Saviotti, P. P. and Metcalfe, J. S. (eds.), *Evolutionary Theories of Economic and Technological Change*, Reading: Harwood Academic Publishers

Saviotti, P. P. and Metcalfe, J. S. (eds.) (1991), *Evolutionary Theories of Economic and Technological Change*, Reading: Harwood Academic Publishers

Saxenian, A. (1991), 'The origins and dynamics of production networks in Silicon Valley', *Research Policy*, 20(5): 423–37

Scherer, F. M. (1965), 'Firm size, market structure, opportunity and the output of patented inventions', *American Economic Review*, 55(5): 1097–123

  (1973), *Industrial Market Structure and Economic Performance*, Chicago: Rand McNally

  (1980), *Industrial Market Structure and Economic Performance*, second edn., Chicago: Rand McNally

  (1982a), 'Inter-industry technology flows in the US', *Research Policy*, 11(4): 227–45

  (1982b), 'Demand pull and technological innovation revisited', *Journal of Industrial Economics*, 30: 215–18

  (1983), 'The propensity to patent', *International Journal of Industrial Organisation*, 1: 107–28

  (1986), *Innovation and Growth. Schumpeterian Perspectives*, Cambridge, Mass.: MIT Press

  (1992), *International High Technology Competition*, Cambridge, Mass., Harvard University Press

Scherer, F. M. and Perlman, M. (eds.) (1992), *Entrepreneurship, Technological Innovation and Economic Growth: Studies in the Schumpeterian Tradition*, Ann Arbor: University of Michigan Press

Schmookler, J. (1966), *Invention and Economic Growth*, Cambridge, Mass.: Harvard University Press

Scholz, L. (1992), 'Innovation surveys and the changing structure of investment in different industries in Germany', *STI Review*, 11: 97–117

Schon, D. A. (1973), 'Product champions for radical new innovations', *Harvard Business Review*, March–April

Schonberger, R. (1982), *Japanese Manufacturing Techniques: Nine Hidden Lessons in Simplicity*, New York: Free Press

Schumpeter, J. A. (1911), *Theorie der wirtschaftlichen Entwicklung*, Berlin: Duncker & Humblot, sixth edn. 1964

  (1912 [1934]), *The Theory of Economic Development*, Cambridge,. Mass.: Harvard University Press

  (1928), 'The instability of capitalism', *Economic Journal*, 38: 361–86

  (1939), *Business Cycles: A Theoretical, Historical and Statistical analysis of the Capitalist Process*, 2 vols, New York: McGraw-Hill

  (1942), *Capitalism, Socialism and Democracy*, New York: McGraw-Hill

Schwitalla, B. (1993), *Messung und Erklärung industrieller Innovationsaktivitäten mit einer empirischen Analyse für die westdeutsche Industrie*, Heidelberg: Physica-Verlag Springer

Sciberras, E. (1977), *Multinational Electronic Companies and National Economic Policies*, Greenwich, CT: JAI Press

Scott, A. J. (1991), 'The aerospace-electronics industrial complex of Southern California: the formative years 1940–1960', *Research Policy*, 20(5): 439–56

Scott, M. F. 1989, *A New View of Economic Growth*, Oxford: Clarendon Press

Senker, J. (1993), 'The role of tacit knowledge in innovation', Brighton: University of Sussex, SPRU, mimeo

Senker, P. J., Swords-Isherwood, N., Brady, T. M. and Huggett, C. M. (1985), 'Maintenance Skills in the Engineering Industry: The Influence of Technological Change', EITB Occasional Paper 8, revised second edn.

Sharp, M. (1985), *The New Biotechnology: European Governments in Search of a Strategy*, Sussex European Paper 15, Brighton: University of Sussex

(1991), 'Pharmaceuticals and biotechnology: perspectives for the European industry', in Freeman, C., Sharp, M. and Walker, W. (eds.), *Technology and the Future of Europe*, London: Pinter

Sharp, M. and Holmes, P. (eds.) (1988), *Strategies for New Technologies*, London: Philip Allan

Sharp, M. and Shearman, C., (1987), *European Technological Collaboration*, London: Routledge and Kegan Paul for the Royal Institute of International Affairs

Shionoya, Y. (1986), 'The science and ideology of Schumpeter', *Rievista Internazionale di Scienze Economiche e Commerciali*, 33(8): 729–62

Silverberg, G. (1984), 'Embodied technical progress in a dynamic economic model: the self-organisation paradigm', in Goodwin, R. M., Krüger, M. and Vercelli, A. (eds.), *Nonlinear Models of Fluctuating Growth*, Berlin, Heidelberg, New York, Tokyo: Springer Verlag

(1987), 'Technical progress, capital accumulation and effective demand: a self-organisation model', in Batten D., Casti, J. and Johansson, B. (eds.), *Economic Evolution and Structural Adjustment*, Berlin, Heidelberg, New York, Tokyo: Springer Verlag

(1988), 'Modelling economic dynamics and technical change', in Dosi, G. *et al.* (eds.), *Technical Change and Economic Theory*, London: Pinter

(1990), 'Adoption and diffusion of technology as a collective evolutionary process', in Freeman, C. and Soete, L. (eds.), *New Explorations in the Economics of Technical Change*, London: Pinter

Silverberg, G., Dosi, G. and Orsenigo, L. (1988), 'Innovation, diversity and diffusion: a self-organising model', *Economic Journal*, 98: 1032–55

Silverberg, G. and Lehnert, D. (1992), 'Long waves and evolutionary change in a simple Schumpeterian model of technical change', paper at MERIT Conference, Maastricht, December

Silverberg, G. and Soete, L. (eds.) (1994), *The Economics of Growth and Technical Change. Technologies, Nations, Agents*, Aldershot, Edward Elgar

Simon, H. A. (1955), 'A behavioural model of rational choice', *Econometrica*, 19: 99–118

(1959), 'Theories of decision-making in economics and behavioral science', *American Economic Review*, 49: 253–83

(1978), 'Rationality as process and as product of thought', *American Economic Review*, 68: 1–16

(1979), 'Rational decision-making in business organisations', *American Economic Review* 69 (4): 493–513

Simon, H. A., Egidi, M., Marris, R. and Viale, R. (1992), *Economics, Bounded Rationality and the Cognitive Revolution*, Aldershot: Edward Elgar

Sirilli, G. (1987), 'Patents and inventors: an empirical study', *Research Policy*, 16(2–4): 157–74

Slaughter, S. (1993), 'Innovation and learning during implementation: a comparison of user and manufacturer innovation', *Research Policy*, 22(1): 81–97

Smith, A. (1776), *Wealth of Nations*, London: Dent (1910)

Smith, H. L., Dickson, K. and Smith S. L. (1991), 'There are two sides to every story: innovation and collaboration within networks of large and small firms', *Research Policy*, 20(5): 457–68

Smith, K. (1991), 'Innovation policy in an evolutionary context', in Saviotti, P. P. and Metcalfe, J. S. (eds.), *Evolutionary Theories of Economic and Technological Change*, Reading: Harwood Academic Publishers

Smith, K. and Vidrei, T. (1992), 'Innovation activity and innovation outputs in Norwegian industry', *STI Review*, 11: 11–35

Smith, S. (1991), 'A computer simulation of economic growth and technical progress in a multi-sectoral economy', in Saviotti, P. P. and Metcalfe, J. S. (eds.), *Evolutionary Theories in Economic and Technological Change*, Reading: Harwood Academic Publishers

Soete, L. (1979), 'Firm size and innovative activity: the evidence reconsidered', *European Economic Review*, 12(4): 319–40

(1981), 'A general test of the technological gap trade theory', *Weltwirtschaftliches Archiv*, 117(4): 638–60

(1987), 'The impact of technological innovation on international trade patterns: the evidence reconsidered', *Research Policy*, 16(2–4): 101–30

(1991), (rapporteur), Synthesis Report, TEP, *Technology in a Changing World*, Paris: OECD

Soete, L. and Verspagen, B. (1993a), 'Convergence and divergence in growth and technical change: an empirical investigation', paper prepared for the AEA conference, Anaheim, CA

(1993b), 'Technology and growth: the complex dynamics of catching up, falling behind, and taking over', in Szirmai, A., van Ark, B. and Pilat, D., *Explaining Economic Growth*, Amsterdam: Elsevier

Soete, L. and Wyatt, S. (1983), 'The use of foreign patenting as an internationally comparable science and technology output indicator', *Scientometrics*, 5(1): 31–54

Soete, L., Verspagen, B., Pavitt, K. and Patel, P. (1989), 'Recent comparative trends in technology indicators in the OECD area', Conference on Science, Technology and Growth, Paris, 5–8 June

Solow, R. (1957), 'Technical change and the aggregate production function', *Review of Economics and Statistics*, 39(August): 312–20

Sorge, A. (1993), 'Introduction to Part IV', in Foray, D. and Freeman, C. (eds.), *Technology and the Wealth of Nations*, London: Pinter

Sorge, A., Campbell, A. and Warner, M. (1990), 'Technological change, product strategies and human resources: defining Anglo-German differences', *Journal of General Management*, 15(3): 39–54

Steele, I. (1991), *Managing Technology: A Strategic View*, New York: McGraw-Hill

Sternberg, E. (1992), *Photonic Technology and Industrial Policy*, New York: State University of New York Press

Stiglitz, J. (1987), 'Learning to learn: localised learning and technological progress', in Dasgupta, P. and Stoneman, P. (eds.), *Economic Policy and Technological Progress*, Cambridge: Cambridge University Press

Stobaugh, R. (1988), *Innovation and Competition: The Global Management of Petrochemical Products*, Boston, Mass.: Harvard Business School Press

Stoneman, P. (1976), *Technological Diffusion and the Computer Revolution*, Oxford: Clarendon Press

(1983), *The Economic Analysis of Technological Change*, Oxford: Oxford University Press

(1987), *The Economic Analysis of Technology Policy*, Oxford: Oxford University Press

Stoneman, P. (ed.) (1995), *Handbook of the Economics of Innovation and Technological Change*, Oxford: Blackwell

Storper, M. and Harrison, B. (1991), 'Flexibility, hierarchy and regional development: the changing structure of industrial production systems and their forms of governance in the 1990s', *Research Policy*, 20(5): 407–22

Summers, R. and Heston, A. (1991), 'The Penn world table (mark V): an expanded set of international comparisons', *Quarterly Journal of Economics*, 106: 327–36

Surrey, A. J. (1973), 'The future growth of nuclear power: Part 1 demand and supply, Part II choices and obstacles', *Energy Policy*, 1(1): 107–29; 1(2): 208–24

Surrey, J. (1992), 'Technical change and productivity growth in the British coal industry 1974–1990', *Technovation*, 12(1): 15–37

Surrey, J. and Thomas, S. (1980), 'World-wide nuclear plant programmes: lessons for technology policy', *Futures*, 12(1): 3–18

Svedberg, R. (1991), *Joseph A Schumpeter: His Life and Work*, Cambridge: Polity Press

Swann, P. L. (ed.) (1992), *New Technology and the Firm*, London: Routledge

Sylos Labini, P. (1962), *Oligopoly and Technical Progress*, Cambridge, Mass.: Harvard University Press

Takeuchi, H. and Nonaka, I. (1986), 'The new product development game', *Harvard Business Review*, January/February: 285–305

Tanaka, M. (1991), 'Government policy and biotechnology in Japan', in Wilks, S. and Wright, M. (eds.), *The Promotion and Regulation of Industry in Japan*, London: Macmillan

Taylor, C. T. and Silberston, Z. A. (1973), *The Economic Impact of the Patent System*, Cambridge University Press

Teece, D. J. (1982), 'Toward an economic theory of the multiproduct firms', *Journal of Economic Behaviour and Organization*, 3(1): 39–63

(1987), 'Profiting from technological innovation: implications for integration, collaboration, licensing and public policy', in Teece, D. J. (ed.), *The Competitive Challenge: Strategies for Industrial Innovation and Renewal*, Cambridge, Mass.: Ballinger

(1988), 'The nature and the structure of firms', in Dosi, G. *et al.* (eds.), *Technical Change and Economic Theory*, London: Pinter

Teece, D. J., Pisano, G. and Shuen, A. (1990), 'Firm capabilities, resources and the concept of strategy', CCC Working Paper 90–8, Berkeley: Center for Research in Management

Teubal, M. (1987), *Innovation, Performance, Learning and Government Policy: Selected Essays*, Madison: University of Wisconsin Press

Teubal, M., Yinnon, T. and Zuscovitch, E. (1991), 'Networks and market creation', *Research Policy*, 20(5): 381–92

Thirlwall, A. (1979), 'The balance of payments constraint as an explanation of international growth rate differences', *Banca Nazionale del Lavoro Quarterly Review*, 6: 45–53

Thomas, G. and Miles, I. (1989), *Telematics in Transition: The Development of New Interactive Services in the UK*, Harlow: Longman

Thomas, S. D. (1988), *The Realities of Nuclear Power: International Economic and Regulatory Experience*, Cambridge University Press

Thwaites, A. T. (1978), 'Technological change, mobile plants and regional development', *Regional Studies*, 12: 455–61

Thwaites, A. T. and Oakey, R. P. (eds.) (1985), *The Regional Impact of Technological Change*, London: Pinter

Tidd, J. (1991), *Flexible Manufacturing Technology and International Competitiveness*, London: Pinter

Tilton, J. (1971), *International Diffusion of Technology: The Case of Semi-Conductors*, Washington, DC: Brookings Institution

Tisdell, C. (1981), *Science and Technology Policy: Priorities of Governments*, London: Chapman and Hall

Townsend, J. (1976), 'Innovation in coal-mining machinery', Occasional Paper no. 3, Brighton: University of Sussex, SPRU

Tylecote, A. (1993), 'Managerial Objectives and Technological Collaboration in National Systems of Innovation: The Role of National Variations in Cultures and Structures', *CRITEC Discussion Paper* No. 2, Sheffield: Sheffield University Management School

Tyson, L., (1992), *Who's Bashing Whom: Trade Conflict in High-Technology Industries*, Institute for International Economics, Washington, DC

UNESCO (1969), *The Measurement of Scientific and Technological Activities*, Paris (1963–1990), *Statistical Yearbooks*, Paris

United States International Trade Commission (1990), 'Identification of US Advanced-Technology Manufacturing Industries for Monitoring and Possible Comprehensive Study', Report to the Committee on Finance, United States Senate, on Investigation No. 332–294. US/TC Publication 2319, Washington

Utterback, J. M. (1979), 'The dynamics of product and process innovation in industry', in Hill, C. and Utterback, J. M. (eds.), *Technological Innovation for a Dynamic Economy*, Oxford: Pergamon

(1993), *Mastering the Dynamics of Innovation*, Boston, Mass.: Harvard Business School Press

Utterback, J. M. and Abernathy, W. J. (1975), 'A dynamic model of product and process innovation', *Omega*, 3(6): 639–56

Utterback, J. M. and Suarez, F. F. (1993), 'Innovation, competition and industry structure, *Research Policy*, 22(1): 1–23

Van de Ven, A. H. *et al.* (1989), *Research on the Management of Innovation*, New York: Harper Row

Van Duijn, J. J. (1983), *The Long Wave in Economic Life*, London: Allen & Unwin

Van Hulst, N., Mulder, N. R. and Soete, L. L. G. (1991), 'Export and technology in manufacturing industry', *Weltwirtschaftliches Archiv*, 127: 246–264

Van Raan, A. (ed.) (1988), *Handbook of Quantitative Studies of Science and Technology*, Amsterdam: Elsevier

Van Vianen, B. G., Moed, H. F. and Van Raan, A. J. F. (1990), 'An exploration of the science base of recent technology', *Research Policy*, 19(1): 61–81

Vernon, R. (1966), 'International investment and international trade in the product cycle', *Quarterly Journal of Economics*, 80: 190–207

Verspagen, B. (1992), 'Endogenous innovation in neo-classical growth models: a survey', *Journal of Macro-Economics*, 14(4): 631–62

(1993a), 'R&D and Productivity: A broad cross-section cross-country look', Maastricht, the Netherlands, MERIT Working Paper 93–007

(1993b), *Uneven Growth Between Interdependent Economies*, Aldershot: Averbury

(1994), 'Technology and growth: the complex dynamics of convergence and divergence', in Silverberg, V. G. and Soete, L. (eds.), *The Economics of Growth and Technical Change*, Aldershot: Edward Elgar

Verspagen, B. and Kleinknecht, A. (1990), 'Demand and innovation: Schmookler re-examined', *Research Policy*, 19: 387–94

Villaschi, A. F. (1992), 'The Brazilian National System of Innovation: opportunities and constraints for transforming technological dependency', D.Phil. thesis, University of London

von Hippel, E. (1978), 'A customer-active paradigm for industrial product idea generation', *Research Policy*, 7: 240–66

(1980), 'The user's role in industrial innovation', in Burton, D. and Goldhar, J. (eds.), *Management of Research and Innovation*, Amsterdam: North-Holland

(1982), 'Appropriability of innovation benefit as a predictor of the source of innovation', *Research Policy*, 11(2), 95–115

(1987), 'Cooperation between rivals: informal know-how trading', *Research Policy*, 16(5): 291–302

(1988), *The Sources of Innovation*, Oxford: Oxford University Press

von Tunzelmann, G. N. (1989), 'Market forces and the evolution of supply in the British telecommunications and electricity supply industries', in Silberston, A. (ed.), *Technology and Economic Progress*, London: Macmillan

Wade, R. (1990), *Governing the Market: Economic Theory and the Role of Goverment in East Asian Industrialisation*, Princeton: Princeton University Press

Walker, W. B. (1977), *Industrial Innovation and International Trading Performance*, Connecticut: JAI Press

(1993), 'From leader to follower: Britain's dwindling technological aspirations', in Nelson, R. R. (ed.), *National Innovation Systems: A Comparative Study*, Oxford: Oxford University Press

Walker, W. B. and Lönnroth, M. (1983a), *Nuclear Power Struggles: Industrial Competition and Proliferation Control*, London: Allen and Unwin

(1983b), 'The viability of the civil nuclear industry', in Somat, J. (ed.), *World Nuclear Energy*, Baltimore: Johns Hopkins Press

Walsh, V. (1984), 'Invention and innovation in the chemical industry: demand pull or discovery push', *Research Policy*, 13: 211–34

(1988), 'Technology and the competitiveness of small countries: a review', in Freeman, C. and Lundvall, B.-Å. (eds.), *Small Countries Facing the Technological Revolution*, London: Pinter

Watanabe, S. (1993), 'Work organisation, technical progress and culture', in Foray, D. and Freeman, C. (eds.), *Technology and the Wealth of Nations*, London: Pinter

Weidlich, W. and Braun, M. (1992), 'The master equation approach to non-linear economics', *Journal of Evolutionary Economics*, 2(3): 233–67

Whiston, T. (1989, 1990), 'Managerial and organisational integration needs arising out of technical change and UK commercial structures', *Technovation*, 9(6), 10(1,2,3): 47–58, 95–118, 143–61

Whittaker, M., Rush, H. and Haywood, W. (1989), 'Technical change in the British clothing industry', Occasional Paper, Brighton: Centre for Business Research

Williamson, O. E. (1975), *Markets and Hierarchies: Analysis and Antitrust Implications. A Study in the Economics of Internal Organisation*, New York: Free Press

(1985), *The Economic Institutions of Capitalism*, New York: Free Press

Winter, S. G. (1964), 'Economic natural selection and the theory of the firm', *Yale Economic Essays*, 4: 225–72

(1971), 'Satisficing, selection and the innovating remnant', *Quarterly Journal of Economics*, 85(2, May): 237–61

(1986a), 'Comments on Arrow and Lucas', *Journal of Economics*, 54: 427–34

(1986b), 'Schumpeterian competition in alternative technological regimes', in Day, R. and Eliasson, G. (eds.), *The Dynamics of Market Economies*, Amsterdam: North-Holland

(1987), 'Knowledge and competence as strategic assets', in Teece, D. J. (ed.), *The Competitive Challenge: Strategies for Industrial Innovation and Renewal*, Cambridge, Mass.: Ballinger

(1988), 'On Coase, competence and the corporation', *Journal of Law, Economics and Organisation*, 4(1): 163–80

(1989), 'Patents in complex contexts: incentives and effectiveness', in Weil, V. and Snapper, J. W. (eds.), *Owning Scientific and Technical Information*, New Brunswick: Rutgers University Press

(1993), 'Patents and welfare in an evolutionary model', *Industrial and Corporate Change*, 2(2): 211–32

Wit, G. R. de (1990), 'The character of technological change and employment in banking', in Freeman, C. and Soete, L. (eds), *New Explorations in the Economics of Technical Change*, London: Pinter

Witt, U. (ed.) (1993), *Evolutionary Economics* (International Library of Critical Writings in Economics), Aldershot: Edward Elgar

Wolff, E.N. and Gittleman, M. (1993), 'The role of education in productivity convergence: does higher education matter?', in Szirmai, E., van Ark, B., and Pilat, D. (eds.), *Explaining Economic Growth*, Amsterdam: Elsevier Science Publishers

Wolff, E.N. and Nadiri, M.I. (1993), 'Spillover effects, linkage structure, and research and development', *Structural Change and Economic Dynamics*, 4: 315–31

Womack, J., Jones, D. and Roos, D. (1990), *The Machine that Changed the World*, New York: Rawson Associates (Macmillan)

World Bank (1986), *World Development Report 1986*, New York: Oxford University Press

(1991), *World Development Report, 1991*, New York: Oxford University Press

Wortmann, M. (1990), 'Multinationals and the internationalisation of R&D: new development in German companies', *Research Policy*, 19: 175–83

Zuscovich, E. (1984), 'Une approche meso-économique du progrès technique: Diffusion de l'innovation et apprentissage industriel', doctoral thesis, Strasbourg: Université Louis Pasteur

Index

Abbot, T. A. 168, 191
Abernathy, W. J. 30, 47–8, 78
Abraham, J. 49
Abramovitz, M. 2, 5, 44, 98
Achilladelis, B. G. 23, 32, 34
Acs, Z. J. 26, 36
Afuah, A. N. 31
Aghion, P. 52
Alchian, A. 21
Alderman, N. 45
Allen, R. C. 52
Amable, B. 52, 204
Amendola, G. 11–12, 23, 141, 152, 155, 162, 173, 181, 185
Ames, E. 6, 31
Amin, M. 45
Amsden, A. 45, 203
Andersen, E. S. 20, 22, 25, 44, 48, 51
Angello, M. M. 20
Antonelli, C. 25, 34, 41, 46, 53
Aoki, M. 28, 33, 53
Arcangeli, F. 40–1
Archibugi, D. 1, 7–10, 13, 33, 49, 122, 129, 132, 145, 162
Arthur, W. B. 22, 39, 48, 53
Arundel, A. 18, 32, 39
Asian Newly Industrialised Countries (NICs) 192, 194, 197, 203–5
    see also Hong Kong, Singapore, South Korea and Taiwan
Atkinson, A. 142
Audretsch, D. B. 36
Australia, 62, 66, 78, 90, 94, 186
Austria 90, 92, 215, 219, 222
Auzeby F. 50
Ayres, R. V. 34, 41

Baba, Y. 26–7, 33–4, 51
Bailey, M. W. 42

Balassa, B. 145, 212
Barras, R. 34–5, 51
Barro, R. J. 105
Basberg, B. 33, 49
basic research 23, 37, 45
Baumol, W. J. 105, 116
Beelen, E. 18
Belgium 92, 117, 180, 215, 222
Bell, M. 29, 55
Bellini, N. 45
Bernal, J. D. 37
Bertin, G. 57
Bessant, J. 33–4, 41–2
Bianchi, P. 45
Bijker, W. E. 37, 43
bio-technology 24–5, 36, 42
Birdzell, L. E. 6
Blackman, B. S. A. 105, 116
Blaug, M. 16
bounded rationality 21
Boyer, R. 42, 44, 48, 52, 204
Bradford, C. 203
Brady, T. M. 29, 35, 51
Branson, W. H. 203
Braun, M. 31, 34
Brazil 45
    see also Latin American countries
Bressand, A. 27, 35
Britain – see United Kingdom (UK)
Brooks, H. 29
Burnell, J. 33
Burns, T. 27
business cycles 16, 18, 91

Callon, M. 39
Camagni, R. 41
Campbell, A. 42
Canada 57, 66, 90, 92, 125, 133, 136, 145, 147, 152–4, 162–3, 215, 219

see also North America; North American
    Free Trade Area (NAFTA)
Cantwell, J. 33, 45, 56, 67, 137, 142–3, 152
capital – accumulation 85
capital – human 79, 105, 107
capital – intangible 55
capital – physical 107
Carlsson, B. 24, 41
Carter, C. F. 23
Cassiolato, J. 31, 34, 51
Casson, M. 45
Cawson, A. 172
Cesaretto, S. 33, 49, 50
Chakrabarti, A. K. 42
Chandler, A. D. 2, 42, 51, 80
Chenery, H. 107
Chesnais, F. 26, 45, 189
Chew, W. 68
Chiaromonte, F. 53
Chow, P. 203
Christensen, J. L. 18, 34, 51
Clark, J. 20, 22, 32, 84
Clark, K. B. 27, 30, 47, 68
classical economists 5
Coase, R. 26
Cobb–Douglas production function 107
Coe, D. T. 99–100, 118
Cohen, W. M. 24, 28, 125
Cohendet, P. M. 25, 46
competition policy 224
Constant Market Share Analysis (CMSA)
    192, 194, 197, 203
consumer behaviour 18
Contribution to Trade Balance (CTB) 145,
    148, 153–5, 162
convergence – economic 132, 137, 142
Coombs, R. 24, 30, 37, 42, 47
Cooper, C. 29
Corbett, J. 78
Cowan, R. 34
Cressey, P. 42
cumulative causation 175
Cusumano, M. A. 35
Cyert, R. M. 21

Dahlman, C. 59
Dahmen, E. 44
Dalum, B. 49, 84
Daniels 178
Dankbaar, B. 34, 42
Darwin, C. 21
Dasgupta, P. 18, 180
data workers 116–17
David, P. A. 4, 22, 39, 40–2, 46, 48, 53,
    180
Davies S. 39, 45

de Broucker, P. 62, 78
De la Mothe, J. 49
DeBresson, C. 27
Deiaco, E. 50
Denison, E. F. 5
Denmark 90, 136, 215, 218
Dertouzos, M. 42
development economics 17
Dickson, K. 26
Dockes, P. 46, 48
Dodgson, M. 2, 24–5, 47
Dollar, D. 123
domestic producers – advanced 211, 223
domestic users – advanced 208, 211, 223
Dore, R. 28, 33
Dosi, G. 2, 7, 18, 20–1, 23, 25, 29–32, 34,
    40–4, 47–9, 51, 53–4, 56, 83–5, 89,
    100, 141, 173, 175–6, 181
Du Pont 30–1
Dunning, J. H. 2, 45

East Asian countries 6, 59, 61, 78
    see also Japan, Korea, Taiwan
Eastern Europe 48
economic development 7–8, 209
Edquist, C. 18, 41
education 7, 61, 78, 105–6
Egidi, M. 21
Eliasson, G. 23, 42, 53
employment 16, 18
Enos, J. L. 34
entrepreneurs 17, 20–1
environmental issues 18
Ergas, H. 18, 32, 56
Ermoliev, Y. M. 53
Ernst, D. 189, 194, 203
Europe 44, 50, 52, 56, 61–2, 68, 78, 84, 139,
    197
    firms 76
    Patent Office 181–2
European Community 139, 147, 168, 194,
    204
European Economic Community (EEC)
    68
European Union, Commission of 178, 180,
    184, 188, 190, 192, 203–5
Evans, P. 24, 38
Evenson, R. E. 118
evolutionary economics 1–2, 17, 21, 42, 83,
    172, 186, 209
evolutionary models – biology 22

Fagerberg, J. 13, 25, 33, 49, 56, 83–4, 89, 99,
    141, 173, 208
Faulkner, W. 25
Fecher, F. 99

Finland 18, 57, 66, 68, 90, 92, 215, 218
    firms 67
    see also Scandinavia
Flamm, K. 34, 39
Fleck, J. 38, 40–1, 43
Foray D. 23, 25–6, 44, 46
Fordism 43–4, 84
Foreign Direct Investment (FDI) 8, 98
Frame, J. 181
France 12, 50, 61, 78, 90, 92, 117–18, 125–7,
    133, 136–9, 145, 147, 152–5, 163,
    184, 215, 219, 222–3
Francois, J.-P. 50
Franko, L. 56
Fransman, M. 26–7
Fraunhofer Institute for Systems and
    Innovation Research (FhG-ISI)
    169
Freeman, C. xi, 1–3, 9, 13, 16, 20, 22–5,
    27–35, 38–9, 43–4, 49, 52–3, 56, 84,
    138, 197, 204
French Regulation School 48
Frenkel, A. 173, 180
Friedman, D. B. 27
Friedman, M. 21
Fujimoto, T. 27, 68
Fuller, J. K. 33–4

Gaffard, J. L. 23
Gann, D. 34
Gardiner, P. 46
Gazis, D. L. 25
Gehrke, B. 169, 173, 181, 185
general equilibrium theory 22
general technologies 26, 36
Germany 6, 12, 29, 50, 56–7, 61–2, 66, 68,
    78–9, 85, 89–92, 117–18, 125, 127,
    129, 133, 136, 138–9, 145, 147,
    154–5, 163, 169, 178, 182, 185, 203,
    215, 219
    training 78
Geroski, P. 56
Gerschenkron, A. 2, 6, 98
Gershuny, J. 42
Gianetti, G. 42, 53, 173
Gibbons, M. 24, 38, 40, 53
Gibbs, D. 45
Gilfillan, S. C. 32
Gilole, B. 41
Gittleman, M. 7, 98, 105
Gjerding, A. N. 42, 46
globalisation 209–10
Goddard, J. B. 45
Gold, B. 39
Golden Age 1, 84
Gomulka 48

Goodwin, R. M. 52
Gordon, R. J. 111
Goto, A. 26–8
Gowing, M. 34
Granstrand, O. 23, 29
Graves, A. 27, 34
Greece 90, 92, 125–6
Greenstein, S. 22, 41, 46
Griliches, Z. 2, 5, 32–3, 49, 56, 105, 117–18
Grossman, I. 52
growth xi, 1–3, 7–8, 13–14, 17, 43, 55, 83–6,
    91–2, 94–5, 99–100, 105–7, 110,
    122–3, 132, 173
    productivity 100
    theory 6, 10, 18, 186
    see also new growth theory
Grubler, A. 39, 41–3
Grupp, H. 10, 12, 23, 28, 33–4, 49, 139,
    168–9, 171, 173, 178, 180–1, 185–6
Guerrieri, P. 10–12, 141, 145, 152, 155, 162,
    188, 190–2, 197, 203

Hagedoorn, J. 26–7, 45
Hahn, F. 22
Hakanson, H. 26
Hamberg, D. 36
Hanusch, H. 17
Hardy, R. 33
Harlow, C. J. E. 34
Harris, R. I. D. 45
Harrison, B. 45
Hartley, J. 168
Hawkins, R. 46
Hayes, R. 78
Haywood, W. 34, 41–2
Heckscher–Ohlin hypothesis 174
Heertje, A. 17–18, 20, 34, 48, 51
Heiner, R. 21
Helpman, E. 52, 99, 118
Heraud, J-A. 46
Herrick, P. 168
Hessen, B. 37
Heston, A. 106
Heston, H. 100
Hicks, D. 45
high technology 168
    industries 188, 197, 204
    exports 190
    externalities 189
    products 191, 197, 204
    taxonomy 191
    trade 190, 203, 205
Hill, C. 47
Hilpert, U. 18
Hirooka, M. 45
Hirschman, A. 2

Hobday, M. 23, 45
Hodgson, G. 21–2, 48, 54
Hoffmaister, A. W. 100, 118
Hoffmann, K. 34
Hofmeyer, O. 28
Holbeck, J. 46
Holland – see Netherlands, the
Hollander, S. 30, 32
Hollingsworth, R. 44
Holmes, P. 18, 172
Hong Kong – see Asian Newly
    Industrialised Countries (NICs)
Hounshell, D. A. 34
Howell, D. R. 116
Howells, J. 45
Howitt, P. 52
Hu, Y-S. 45, 78
Hufbauer, G. C. 32, 49, 153
Huggett, C. M. 29
Hughes, K. 28
Hughes, T. P. 41
Hungary 62

IKE data base – University of Aalborg 215
Imai, K. 25–7, 33
Information and Communication
    Technology (ICT) 24–7, 35, 42, 52
Information Technology (IT) 110–11, 117,
    119, 186
information workers 116–17
innovation – demand-pull 21, 37–8, 47
    see also technology – demand-pull
innovation – diffusion 17–18, 20, 22, 39–41,
    43, 46, 53, 84
innovation – incremental 22, 30–3, 35, 37,
    40
innovation – measurement 49
innovation – organisational 42, 51
innovation – policy 18, 44
innovation – radical 22, 30–5, 37, 40, 42,
    47
innovation – statistics 32
innovation – supply/technology push 21,
    37–8, 47
    see also technology – supply-push
innovation – taxonomies 35
institutional economics 1–2, 4, 7
institutional failures 77
intellectual property rights 3, 181, 184, 186
international specialisation xii
International Trade Administration (ITA)
    191
Ireland 57, 62, 90, 92
Irvine, J. 49
Isard, P. 45
Italy 6, 50, 62, 90, 92, 133, 136–7, 139,

144–5, 147, 153–5, 162–3, 169, 178,
    215, 219
Itami, H. 33
Iwai, K. 53

J-firm 28, 51–2
    see also Japan – firms
Jacobsson, S. 24, 34, 41
Jagger, N. S. B. 41–2
Jaikumar, R. 34
Jang-Sup Shin 45
Japan 6, 9, 44, 52, 56–7, 61–2, 66, 68, 79,
    84–5, 89–92, 94, 125–7, 129, 136–8,
    145, 147, 154–5, 163, 168, 180–2,
    185, 188, 190, 192, 194, 197, 204–5,
    215, 218
    firms 23, 26–8, 33, 51, 66–8, 194, 197
        see also J-firm
    training 78
    see also East Asian countries
Jevons, F. 24, 38
Jewkes, J. 16–17
Johansson, J. 26
Johnson, B. 42–4
Johnson, C. 197
Johnston, R. 24, 168
Jones, D. 28, 34, 42
Jorgenson, D. 5

Kalacheck, E. D. 18
Kaldor, N. 2, 13, 175
Kaldorian economics 52
Kallehauge, L. 42
Kalypso, N. 27, 35
Kamien, M. 21, 36
Kaniovski, Y. M. 53
Kaplinsky, R. M. 29, 36
Katz, B. G. 34, 37
Kay, N. 35
Keck, O. 34
Kelley, M. B. 29
Kellman, H. M. 203
Kennedy, C. 16, 19
Keynes, J. M. 22
Keynesian economics 5, 52
Kitti, C. 57
Klein, B. H. 42–3
Kleinknecht, A. 25–6, 32, 37
Klevorick, A. K. 4, 26, 28, 46
Kline, S. 39
Knowledge – accumulation 23
    national patterns of 62
    see also technology – accumulation
knowledge – intensity 49, 123
knowledge – spillovers 118
knowledge workers 116–17

Kodama, F. 27, 41, 168, 188, 197
Korea 45
    see also East Asian countries
Koschatsky, K. 173, 180
Krauch, H. 18
Kreinin, M. E. 197
Krugman, P. xi, 11, 49, 152–3, 163, 172,
    189, 192
Krupp, H. 34
Kuhn, T. 2, 43, 47
Kuznets, S. 98

Lall, S. 98
Landes, M. 2, 42
Langrish, J. 24, 38
Lastres, H. 25
Latin America 6, 59
    see also Brazil
Law, J. 37, 43
Lawrence, P. 18, 197
Lazonick, W. 42
Legler, H. 168–9, 173, 181, 203
Lehnert, D. 53
Leroux-Demers, T. 23
Lester, R. 42
Levin, R. C. 4, 24, 26, 28, 46
Levine, R. 105
Levinthal, D. A. 28, 125
Lewis, A. 2
Lichtenberg, F. R. 100, 107, 117
Limpens, I. 18
Lines, M. 23, 32, 34
List, F. 2, 208
Lonroth, M. 34
Lorenz, D. 173
Lovio, R. 34, 36
Lower Saxony Institute for Economic
    Research (NIW) 169
Lucas, R. 21
Lundgren, A. 4, 36, 52
Lundvall, B-A. 4, 13, 25, 27, 31–2, 42, 44,
    54, 83, 85, 138, 209

MacDonald, S. 31, 34
Machin, S. 56
Machlup, F. 31
Mackenzie, D. 37, 43
MacKerron, G. S. 34
MacQueen, D. H. 36
Madsen, P. T. 42
Magee, S. P. 192
Mahajan, V. 39
Maital, S. 173, 180
Malerba, F. 34
Manchester Study of Queen's Awards, the
    24
Mankiw, N. G. 105, 107

Mansell, R. 41–2, 46, 51
Mansfield, E. 2, 4, 23–4, 26–7, 34–5, 39–41,
    46
March, J. G. 21
Marquis, D. G. 37–8
Marris, R. 21
Marsden, D. 29
Marshall, A. 29, 51
Martin, B. R. 49
Marx, K. xi, 2, 16–17
Marxist economics 1
Matthews, J. 42
Mayer, C. 78
McDougal, G. 62, 78
McGuehin, R. 168
McMullen, K. 62, 78
Menger, C. 54
Mensch, G. 31–2
Metcalfe, J. S. 22, 39–40, 47, 53
Meyer-Krahmer, F. 18, 50
Mexico – *see* North American Free Trade
    Area (NAFTA)
Michie, J. 1, 13
Midgeley, D. F. 39–40
Milana, C. 10, 145, 188, 190–2
Miles, I. 41–2, 51
Miller, R. 23
Mirrlees, J. A. 175
Mjoset, L. 25, 44
Moed, H. F. 25
Moggi, M. 41
Mohnen, P. 189
Mollina, A. H. 25, 34, 37
Morgan, K. 41, 45–6
Morris, P. J. T. 34
Morrison, P. D. 39–40
Morton, J. A. 28
Mowery, D. C. 18, 24, 26, 28, 38, 194
Mulder, K. F. 47
Mulder, N. R. 180
multi-national companies (MNCs) 36, 45,
    51, 138, 180, 210
Munt, G. 10, 12, 168
Myers, S. 37–8

Nadiri, M. I. 117
Nakicenovic, N. 39, 41–2
Narin, F. 25, 33
Nasbeth, L. 41
National Institute of Economic and Social
    Research 61
national systems of innovation 25, 32, 44–5,
    48, 77, 95, 83, 124, 138, 176, 185
    dynamic 77, 80
    myopic 77–8, 80
    see also research institutes
natural selection 22

Nelson, R. 2, 4–6, 11, 13, 18–19, 23–6, 28,
    30, 34–5, 43–7, 50–1, 53, 56, 79, 83,
    105, 138, 192
neoclassical economics 5, 7, 18, 83–4, 174,
    208, 222
Netherlands, the 12, 50, 57, 79, 90, 94, 127,
    136, 145–7, 152–4, 162–3, 169, 185,
    215, 219
new growth theory xi, 11, 45, 52, 83, 172
    see also growth theory
new materials technology 24–5
new trade theory xi, 172
Newton, K. 62, 78
Nietsche, F. H. 20
Niosi, J. 44–5
Noma, E. 25, 33
Nonaka, I. 27
Norfolk, L. 168
North America 78
    see also USA, Canada
North American Free Trade Area
    (NAFTA) 178
    see also Mexico, United States of
    America, Canada
Norway, 50, 66, 68, 90, 92, 215, 218
    see also Scandinavia

O'Connor, D. 189, 194, 203
Oakey, R. 36, 45
office, computer and accounting equipment
    111
Ohmae, K. 28, 45
Okimoto, D. I. 197
Oldham, G. 29
oligopoly 35
Olivastro, D. 25
Olleros, X. 23
Olson, M. 6, 48
Organisation for Economic Co-operation
    and Development (OECD) 8, 29, 40,
    56–7, 62, 80, 85–8, 90, 92, 94,
    99–100, 117–18, 123, 125, 128–9,
    133, 137, 143, 168–9, 178, 180–1,
    183, 185–6, 192, 208, 215
    Directorate for Science, Technology and
        Industry 49
    Frascati Manual 29
    OSLO Manual 50
    Trade series 215
    publications 18, 26, 30, 32, 39, 42, 49–50,
        188–9
Orsenigo, L. 25, 42, 53
Ostry, S. 189

Padoan, P. C. 11–12, 141, 152,155, 162
Papagni, E. 173
Pasinetti, L. L. 18

Patel, P. 7–10, 33, 45, 49, 55–6, 61, 66, 180,
    204
patents 59, 85–7, 90, 92, 124, 129–31, 133,
    136, 144, 177, 180–2, 184
    statistics 33, 35
    US data 56, 68, 100, 123, 132, 143–4, 182,
        186
    index of 100
path dependence 22
Paulinyi, A. 33
Pavitt, K. 2, 4, 7–10, 11, 13, 18, 23–4, 29–33,
    35–6, 45, 49, 54–6, 61, 66, 83–5, 89,
    100, 141–3, 146, 163, 173, 180–1, 204
Peacock, T. 49
Pearce, R. D. 45
Peck, M. J. 18
Penrose, E. 2, 22, 29
Perelman, S. 99
Perez, C. 31–2, 47–8
Perlman, M. 17
Perrin, J. 44
Perucci, A. 169, 181, 185
Peterson, R. A. 39
Petit, P. 34, 53, 117
Petroski, H. 33
Phelps, E. S. 105
Phillips, A. 20, 34, 37
Pianta, M. 7–10, 33, 49, 83, 122, 129, 132,
    145
Piore, M. J. 42, 110
Pisano, G. 23, 47
Polanyi, M. 2
Poon, A. 34–5
Porter, M. 13, 45, 68, 77, 168, 189, 208–12,
    216, 221–3
Portugal 57, 62, 90, 92, 125–6
Posner, M. 9, 11, 49, 56, 153, 173
Posthuma, A. 34
Prais, S. J. 29, 61, 78
Price, D. de S. 2, 23
production function approach 18
protectionism 208, 222
public good 5, 107, 180
public policy 15

Quinn, J. B. 34
Quintas, P. 35, 51
Quintella, R. 34, 47

Ray, G. F. 41
Reekie, W. D. 33
regional development 18
Reich, L. S. 28
Reich, R. B. 45
Reijnen, J. O. N. 25–6
Renelt, D. 105
representative agent 20, 20

Research and Development (R&D) 5, 8, 10,
    12, 21, 26–31, 34–6, 44–5, 49–50,
    56–7, 66, 68, 77–8, 84, 86, 88, 90,
    92, 95, 98–101, 110–11, 117–19,
    123, 126, 128–9, 131, 138, 147,
    152, 168, 174, 176–8, 185, 188,
    190–1, 194
  convergence 57
  expenditure 85–7, 100, 104–7, 124–5,
    168–9, 171
  intensity 35, 89–92, 94, 100, 105, 110,
    127, 129, 168–9, 171, 185
  measure 168–9
  slipovers 119
  statistics 30, 34, 49
research institutes 2, 28, 43–4
  see also national system of innovations
Revealed Comparative Advantage (RCA)
    145, 147–8, 152–4, 162, 212, 215
Revealed Technical Advantage (RTA) 62
Revealed Technology Comparative
    Advantage (RTCA) 145, 147–8,
    154–5, 162
reverse engineering 26, 28
Ricardo, D. 11
Richards, A. 24, 47
Roberts, E. B. 23
Roberts, J. H. 39–40
Robinson, J. 1
Robinson, S. 107
Robson, M. 32, 36, 146
Rogers, E. 17
Romeo, A. A. 41
Romer, P. 52, 105, 107
Roobeek, A. J. M. 48
Roos, D. 28, 34, 42
Rosario, M. 46
Rosenberg, N. 2, 6, 17, 23, 34, 38–9, 41, 53,
    142, 194
Ross-Larsen, B. 59
Rothwell, R. 2, 18, 24, 26–7, 31, 39, 46
Rush, H. 34, 42
Russo, M. 45
Ruttan, V. 22

Sabel, C. F. 27, 110
Sahal, D. 47–8, 142
Sako, M. 25–7
Sakyarakwit, W. 29
Salomon, J. J. 18
Salter, W. 29
Samuels, R. J, 27, 34
Santarelli, E. 162
Sante Fe Institute 22
SAPPHO project 24–5, 27
Saviotti, P. P. 22, 24, 30, 37, 42, 44

Sawers, D. 16–17
Saxenian, A. 45
Sayer, A. 45
Scandinavia 78
  see also Norway, Sweden, Finland
Schakenraad, J. 26–7
Schasse, U. 169, 173, 181, 185
Scherer, F. M. 2, 17, 21, 33–7, 133, 188–9
Schiffel, D. 57
Schmidt, S. K. 46
Schmookler, J. 33, 37–8
Schneider, V. 51
Scholz, L. 50
Schon, D. A. 22
Schonberger, R. 27
Schumpeter, J. 2, 16–17, 20, 22, 30, 35, 39,
    42, 47–8, 51, 173
Schumpterian economics xii, 1, 5, 83, 224
Schwartz, N. 21, 36, 46
Schwartzkopf, A. 23, 32, 34
Schweitzer, T. 62, 78
Schwitalla, B. 169
Sciberras, E. 34
Scott, A. J. 45
Second World War 32, 79, 192
Senker, P. J. 29, 30
Sercovitch, F. 29
Sharp, M. 18, 24, 204
Shearman, C. 204
Shionoya, Y. 20
Shuen, A. 23, 47
SIE World Trade Data Base 192
Siedule, T. 62, 78
Siegel, A. 117
Silberston, Z. A. 46
Silverberg, G. 52–3
Simon, H. A. 21
Singapore – see Asian Newly Industrialised
    Countries (NICs)
Singh, S. 45
Single European Market 172
Sirilli, G. 33, 49–50
Sjolander, S. 23
Slaughter, S. 25, 31
Smith, A. 2, 31
Smith, H. L. 26
Smith, J. K. 34
Smith, K. 50, 53
Smith, S. 18
Soete, L. 2, 18, 20–1, 32–4, 39, 49, 53–4, 56,
    83–5, 89, 100, 123, 137, 141, 143,
    173, 175–6, 180–1, 186
Soligny, P. 18
Solow, R. 5, 42
Solow residual 5, 18
Sorge, A. 42–3

South Korea 59, 99
  see also Asian Newly Industrialised
    Countries (NICs)
Spain 57, 90, 94, 132, 178, 215, 219
Stalker, G. M. 27
Standard and Poor 169
Stankiewicz, R. 41
Steele, I. 24
Steinmuller, E. 22, 41
Sternberg, E. 34
Stiglitz, J. 23, 25, 142
Stillerman, R. 16–17
Stobaugh, R. 30, 34
Stoneman, P. 2, 18, 39, 41, 53
Storper, M. 45
Suarez, F. F. 36, 46
Summers, R. 100, 106
Surrey, J. 32, 34
Svedberg, R. 20
Swann, P. L. 23
Sweden 50, 52–3, 57, 79, 90, 94, 117, 125,
    127, 136–7, 146–7, 152–4, 162–3,
    169, 215, 219
  see also Scandinavia
Sweden – firms 68
Switzerland 57, 61, 66, 68, 79, 94, 125, 127,
    133, 136–7, 146–8, 154, 162–3,
    184–5, 215, 218
Swords-Isherwood, N. 29
Sylos Labini, P. 21, 44
Syrquin, M. 107

tacit knowledge 4, 29–30, 45, 55
Tahar, G. 34, 53
Taiwan 45, 59
  see also East Asian countries; Asian
    Newly Industrialised Countries
    (NICs)
Takai, S. 34, 51
Takeuchi, H. 27
Tanaka, M. 27
Taylor, C. T. 46
technical change xi–xii, 1–3, 16, 18, 20,
    175
  measurement 49
technical collaboration 26
technology – accumulation 55, 66, 77–8
  see also knowledge – accumulation
technology – advantage 142
technology – 'black box' approach 16–17,
    31
technology – capability 98, 122, 141, 189
technology – change 223
technology – competence 6–7, 9, 11
technology – convergence xii, 10, 122–3,
    125, 139

technology – demand-pull 78
  see also innovations – demand-pull
technology – diffusion 123, 139
technology – distance – measure of 123,
    133, 136
technology – gap 9, 11, 83, 137, 153, 163,
    175
technology – generic 42, 52
technology – globalisation 25
technology – high level 12
technology – measure 56, 100
technology – national specialisation
    patterns 221
technology – policy 186
technology – proprietary nature 4
technology – specialisation 66, 123, 141–4,
    146–8, 152–5, 162–3, 173
technology – supply-push 78
  see also innovations – supply-push
technology – trajectory 19, 35, 38, 41, 43,
    47, 124, 142, 152
technology – transfer 124, 138
technology – user specific 210
Teece, D. J. 23, 29, 46–7
Teubal, M. 18, 23–4, 26
Thirlwall, A. P. 16, 19
Thomas, G. 41–2, 51
Thomas, S. 34
Thwaites, A. 45
Tidd, J. 34
Tisdell, C. 18
Toninelli, P. A. 42, 53, 173
Torrisi, F. 34
total factor productivity (TFP) 99–100,
    118
Townsend, J. 30, 32–3, 36, 146
trade xi, 1–3, 12, 14, 17–18, 49, 141, 143,
    173–6, 181, 189, 197
  specialisation 141–4, 146–8, 152–5,
    162–3, 173
  indicator 145
  national patterns 141
  theory 174, 208, 222
transaction costs 27–8, 209
Turner, K. 42
Tyson, L. 85, 189, 197

underdevelopment 18
United Kingdom (UK) 6, 12, 29, 56–7,
    61–2, 66, 78–9, 90, 92, 117–18,
    125, 127, 133, 136–9, 145, 147,
    152–4, 162–3, 184, 204, 215, 219,
    222–3
  firms 67–8
United Nations (UN) 144, 192
  see also UNESCO

United Nations Economic, Scientific and
    Cultural Organisation (UNESCO)
    29, 49, 100
    see also UN
United Statges of America (USA) 5–6, 9,
    12, 44, 56–7, 59, 62, 66, 68, 78–9,
    84–5, 89–92, 94, 116–18, 125, 127,
    129, 132–3, 136–8, 143–5, 152–4,
    168–9, 178, 180–1, 185–6, 188, 190,
    194, 197, 203–5, 215, 219
    Department of Commerce 168
    firms 26–8, 37, 52, 76, 192
    International Trade Commission 168
    National Science Foundation 24, 49
    Project Traces 24–5
    see also North America; North American
        Free Trade Area (NAFTA)
universities 3, 24
user-producer relationships 209–10
Utterback, J. M. 30–2, 35–6, 43, 46–8

Van de Ven, A. H. 24
van Hulst, N. 180
Van Raan, A. J. F. 25, 180
van Reenen, J. 56
Van Vianen, B. G. 25
Vergragt, P. J. 47
Vernon, R. 49, 56
Verspagen, B. 18, 37, 44, 49, 56, 99, 123, 137
Viale, R. 21
Vidrei, T. 50
Villaschi, A. F. 45
virtuous circle 85, 91–2, 95, 175, 186
von Hippel, E. 13, 25–6, 31
von Tunzelman, G. N. 33–4

Wade, R. 45, 203
Wagner, K. 29

Walker, W. B. 18, 34, 45, 49, 204
Wallmark, J. T. 36
Walras, L. 51
Walrasian theory 22
Walsh, V. 24, 30, 34, 37–8, 42, 136
Warner, M. 42
Watanabe, S. 42
Watson, T. J. 37
Webb, S. 33
Weidlich, W. 53
Weil, D. 105, 107
Westphal, L. 59
Whiston, T. 42
Whittaker, M. 34
Williams, B. R. 23, 42
Williamson, O. 26, 28
Winter, S. G. 4, 11, 19, 23, 26, 28–30, 34–5,
    43, 46–7, 50–1, 53–4
Witt, U, 51
Wolff, E. N. 7, 98, 105, 116, 123
Womack, J. 28, 34, 42
World Bank 49, 100
    publication 45
World War Two – see Second World War
Wortmann, M. 45
Wright, G. 6, 79
Wyatt, S. 33, 49, 57
Wynarczyk, W. 45

Yale University Survey 24, 46
Yinnon, T. 24, 26
Young, A. J. 33
Yugoslavia 99

Zegveld, W. 18
Zuscovitch, E. 24, 26, 39, 46
Zysman, J. 197

Printed in the United States
By Bookmasters